Microc

Dogan Ibrahim
Ahmet Ibrahim

Microcontroller Based GSM/GPRS Projects

Advanced Microcontroller Projects

VDM Verlag Dr. Müller

Impressum/Imprint (nur für Deutschland/ only for Germany)

Bibliografische Information der Deutschen Nationalbibliothek: Die Deutsche Nationalbibliothek verzeichnet diese Publikation in der Deutschen Nationalbibliografie; detaillierte bibliografische Daten sind im Internet über http://dnb.d-nb.de abrufbar.

Alle in diesem Buch genannten Marken und Produktnamen unterliegen warenzeichen-, marken- oder patentrechtlichem Schutz bzw. sind Warenzeichen oder eingetragene Warenzeichen der jeweiligen Inhaber. Die Wiedergabe von Marken, Produktnamen, Gebrauchsnamen, Handelsnamen, Warenbezeichnungen u.s.w. in diesem Werk berechtigt auch ohne besondere Kennzeichnung nicht zu der Annahme, dass solche Namen im Sinne der Warenzeichen- und Markenschutzgesetzgebung als frei zu betrachten wären und daher von jedermann benutzt werden dürften.

Coverbild: www.ingimage.com

Verlag: VDM Verlag Dr. Müller Aktiengesellschaft & Co. KG
Dudweiler Landstr. 99, 66123 Saarbrücken, Deutschland
Telefon +49 681 9100-698, Telefax +49 681 9100-988
Email: info@vdm-verlag.de

Herstellung in Deutschland:
Schaltungsdienst Lange o.H.G., Berlin
Books on Demand GmbH, Norderstedt
Reha GmbH, Saarbrücken
Amazon Distribution GmbH, Leipzig
ISBN: 978-3-639-24910-1

Imprint (only for USA, GB)

Bibliographic information published by the Deutsche Nationalbibliothek: The Deutsche Nationalbibliothek lists this publication in the Deutsche Nationalbibliografie; detailed bibliographic data are available in the Internet at http://dnb.d-nb.de.

Any brand names and product names mentioned in this book are subject to trademark, brand or patent protection and are trademarks or registered trademarks of their respective holders. The use of brand names, product names, common names, trade names, product descriptions etc. even without a particular marking in this works is in no way to be construed to mean that such names may be regarded as unrestricted in respect of trademark and brand protection legislation and could thus be used by anyone.

Cover image: www.ingimage.com

Publisher: VDM Verlag Dr. Müller Aktiengesellschaft & Co. KG
Dudweiler Landstr. 99, 66123 Saarbrücken, Germany
Phone +49 681 9100-698, Fax +49 681 9100-988
Email: info@vdm-publishing.com

Printed in the U.S.A.
Printed in the U.K. by (see last page)
ISBN: 978-3-639-24910-1

Copyright © 2010 by the author and VDM Verlag Dr. Müller Aktiengesellschaft & Co. KG and licensors
All rights reserved. Saarbrücken 2010

PREFACE

This book is aimed for the people who may want to learn how to use PIC series of microcontrollers and GSM/GPRS modems in their projects using the mikroC programming language.

The highly popular PIC16F887 microcontroller has been taken and used as an example microcontroller in the book. In addition, the SIM340Z GSM/GPRS modem card is used in modem based projects.

All of the projects in the book have been built and tested. The PIC microcontroller development board EasyPIC 5 is used to construct the projects. In addition, the Smart GSM/GPRS development board is used together with a PC or with the EasyPIC 5 development board to construct modem based projects.

The book makes an introduction to the PIC16F887 microcontroller and then teaches the basic principles of program development using the mikroC programming language. The book then describes how to build simple PIC microcontroller based projects with and without using the modem card.

We hope you enjoy reading the book as much as we enjoyed writing it.

Ahmet Ibrahim
Dogan Ibrahim

London, 2010

TABLE OF CONTENTS

PREFACE		1
TABLE OF CONTENTS		2
LIST OF TABLES		7
LIST OF FIGURES		8

1. **INTRODUCTION** — 11
    1.1 Overview — 11
    1.2 Stand-alone Data Loggers — 12
    1.3 Data Capturing Data Loggers — 12

2. **MICROCONTROLLERS** — 15
    2.1 Overview — 15
    2.2 Microcontroller Systems — 15
        2.2.1 RAM — 17
        2.2.2 ROM — 17
        2.2.3 PROM — 17
        2.2.4 EPROM — 17
        2.2.5 EEPROM — 18
        2.2.6 Flash EEPROM — 18
    2.3 Microcontroller Features — 18
        2.3.1 Supply Voltage — 18
        2.3.2 The Clock — 19
        2.3.3 Timers — 19
        2.3.4 Watchdog — 20
        2.3.5 Reset Input — 20
        2.3.6 Interrupts — 20
        2.3.7 Brown-out Detector — 20
        2.3.8 Analog-to-digital Converter — 21
        2.3.9 Serial Input-output — 21
        2.3.10 EEPROM Data Memory — 22
        2.3.11 LCD Drivers — 22
        2.3.12 Analog Comparator — 22
        2.3.13 Real-time Clock — 22
        2.3.14 Sleep Mode — 22
        2.3.15 Power-on Reset — 23
        2.3.16 Low Power Operation — 23

		2.3.17 Current Sink/Source Capability	23
		2.3.18 USB Interface	23
		2.3.19 Motor Control Interface	23
		2.3.20 CAN Interface	24
		2.3.21 Ethernet Interface	24
		2.3.22 ZigBee Interface	24
	2.4	Microcontroller Architectures	24
		2.4.1 RISC and CISC	25
3.	**PIC 16F887 MICROCONTROLLER**		**26**
	3.1	Overview	26
	3.2	The Specifications of the PIC16F887 Microcontroller	26
	3.3	The Architecture of the PIC16F887 Microcontroller	27
	3.4	Clock Configurations	29
	3.5	Reset Configurations	32
		3.5.1 Power-on Reset	33
		3.5.2 External Reset	34
	3.6	Minimum Crystal Based Configuration	34
4.	**mikroC LANGUAGE FOR PIC MICROCONTROLLERS**		**35**
	4.1	Overview	35
	4.2	Structure of a mikroC Program	36
		4.2.1 Comments	37
		4.2.2 Beginning and Ending of a Program	37
		4.2.3 Terminating Program Statements	38
		4.2.4 White Spaces	38
		4.2.5 Case Sensitivity	38
		4.2.6 Variable Names	39
		4.2.7 Variable Types	40
		4.2.8 Constants	42
		4.2.9 Escape Sequences	44
		4.2.10 Enumerated Variables	45
		4.2.11 Arrays	46
		4.2.12 Pointers	47
		4.2.13 Structures	49
		4.2.14 Operators in C	53
		4.2.15 Modifying the Flow of Control	57
	4.3	PIC Microcontroller Input-Output Programming	67
	4.4	Programming Examples	67
5.	**THE EasyPIC 5 MICROCONTROLLER DEVELOPMENT BOARD**		**72**
	5.1	Overview	72

5.2		Specifications of the EasyPIC 5 Development Board	73
5.3		Simple Projects With the EasyPIC 5 Development Board	73
	5.3.1	BEGIN-END	74
	5.3.2	Statements	74
	5.3.3	IF-ELSE-ENDIF	75
	5.3.4	REPEAT-UNTIL	75
	5.3.5	DO FOREVER – ENDDO	76
	5.3.6	Do n TIMES – ENDDO	76
	5.3.7	WHILE – WEND	76
5.4		PROJECT 1 – FLASHING LEDs	78
	5.4.1	Description	78
	5.4.2	Block Diagram	78
	5.4.3	Circuit Diagram	78
	5.4.4	PDL of the Project	79
	5.4.5	Program Listing	80
	5.4.6	Suggestions for Future Work	86
5.5		PROJECT 2 – MOVING LEDs	87
	5.5.1	Description	87
	5.5.2	Block Diagram	87
	5.5.3	Circuit Diagram	87
	5.5.4	PDL of the Project	87
	5.5.5	Program Listing	88
	5.5.6	Suggestions for Future Work	89
5.6		PROJECT 3 – LED WITH PUSH BUTTON SWITCH	90
	5.6.1	Description	90
	5.6.2	Block Diagram	90
	5.6.3	Circuit Diagram	90
	5.6.4	PDL of the Project	90
	5.6.5	Program Listing	92
	5.6.6	Suggestions for Future Work	93
5.7		PROJECT 4 – COUNTING LCD	94
	5.7.1	Description	94
	5.7.2	Block Diagram	94
	5.7.3	Circuit Diagram	97
	5.7.4	PDL of the Project	98
	5.7.5	Program Listing	99
	5.7.6	Suggestions for Future Work	101
5.8		PROJECT 5 – SENDING DATA TO A PC USING THE RS232 PORT	102
	5.8.1	Description	102
	5.8.2	Block Diagram	102
	5.8.3	Circuit Diagram	105
	5.8.4	PDL of the Project	105

		5.8.5 Program Listing	106
		5.8.6 Testing the Program	109
		5.8.7 Suggestions for Future Work	112
6.	**THE GSM/GPRS MODEM**		**113**
	6.1	Overview	113
	6.2	The GSM System	113
	6.3	The SIM Card	115
	6.4	The Smart GSM/GPRS Development Board	117
	6.5	Using the Smart GSM/GPRS Board with a PC	120
		6.5.1 Setting-up the Hardware	120
		6.5.2 Setting-up the Software	123
		6.5.3 AT Commands	124
	6.6	GSM/GPRS – PC PROJECT 1 – SENDING SMS MESSAGES	128
	6.7	GSM/GPRS – PC PROJECT 2 – SENDING SMS MESSAGES IN PDU MODE	134
	6.8	GSM/GPRS – PC PROJECT 3 – CALLING A MOBILE PHONE	140
	6.9	GSM/GPRS – PC PROJECT 4 – ANSWERING A PHONE CALL	143
	6.10	GSM/GPRS – PC PROJECT 5 – SOME OTHER USEFUL COMMANDS	145
		6.10.1 Command AT+CLTS	145
		6.10.2 Command AT+CBAND	145
		6.10.3 Command AT+CIMI	146
		6.10.4 Command AT+CNUM	146
		6.10.5 Command AT+CRSL	146
		6.10.6 Command AT+CLVL	147
		6.10.7 Command AT+CMUT	147
7.	**USING THE GSM/GPRS MODEM WITH MICROCONTROLLERS**		**148**
	7.1	Overview	148
	7.2	PROJECT 1 – Sending a SMS Text Message	148
	7.3	PROJECT 2 – Sending the Temperature as SMS	156
		7.3.1 Overview	156
		7.3.2 The Temperature Sensor	156
		7.3.3 The Circuit Diagram	157
		7.3.4 The Software	158
	7.4	PROJECT 3 – SENDING SMS USING A MICROCONTROLLER AND A PC	166

8.	**THE REAL TIME CLOCK CHIP**		176
	8.1 Overview		176
	8.2 The PCF8583 RTC Chip		176
		8.2.1 The Control and Status Register	177
		8.2.2 The Counter Register	177
		8.2.3 Format of the Hours Register	178
		8.2.4 Format of the Year/Date Register	179
		8.2.5 Format of the Weekdays/Month Register	180
	8.3 Using the PCF8583 RTC Chip in Microcontroller Projects		180
		8.3.1 Writing to the PCF8583 RTC Chip	180
		8.3.2 Reading From the PCF8583 RTC Chip	182
	8.4 Example Reading From the RTC Chip		182
	8.5 Example Writing to the RTC Chip		184

REFERENCES 185

APPENDIX A – THE ASCII TABLE 186

APPENDIX B - SIM340 MODEM PIN ASSIGNMENTS AND PIN DEFINITIONS 187

APPENDIX C – SOME COMMONLY USED AT COMMANDS 189

APPENDIX D – SMS DATA CODING SCHEME 190

APPENDIX E – SMS DEFAULT ALPHABET 192

INDEX 193

LIST OF TABLES

TABLE 4.1 mikroC Reserved Names
TABLE 4.2 mikroC Variable Types
TABLE 4.3 mikroC Arithmetic Operators
TABLE 4.4 mikroC Relational Operators
TABLE 4.5 mikroC Logical Operators

TABLE 5.1 Pin configuration of HD44780 LCD module
TABLE 5.2 Minimum Required Pins for Serial Communication

TABLE 6.1 GSM Operators in the UK
TABLE 6.2 LED Showing the Modem Status
TABLE 6.3 Some General Purpose AT Commands
TABLE 6.4 Important SIM340Z Modem SMS Commands

LIST OF FIGURES

FIGURE 1.1 Block Diagram of a typical GSM/GPRS and microcontroller based system

FIGURE 2.1 Von Neumann and Harvard Architectures

FIGURE 3.1 Architecture of the PIC16F887 Microcontroller
FIGURE 3.2 RC Clock Mode
FIGURE 3.3 RCIO Clock Mode
FIGURE 3.4 Using a Rezonator
FIGURE 3.5 Using a Crystal
FIGURE 3.6 Using External Clock
FIGURE 3.7 Power-on Reset
FIGURE 3.8 Power-on Reset For Slow Rising Vdd
FIGURE 3.9 External Reset Circuit
FIGURE 3.10 Minimum Crystal Based Configuration

FIGURE 4.1 Structure of a Simple mikroC Program

FIGURE 5.1 EasyPIC 5 microcontroller development board
FIGURE 5.2 Block Diagram of the Project
FIGURE 5.3 Circuit Diagram of the Project
FIGURE 5.4 PDL of the Project
FIGURE 5.5 Program Listing of the Project
FIGURE 5.6 Select the Device and Device Flags
FIGURE 5.7 Write the Program
FIGURE 5.8 Compiler the Program
FIGURE 5.9 Jumper and LED Positions on EasyPIC 5
FIGURE 5.10 Modified Program
FIGURE 5.11 Using define Statement for the Delay Routine
FIGURE 5.12 Using a Function to Introduce Delay
FIGURE 5.13 PDL of the Project
FIGURE 5.14 Program Listing
FIGURE 5.15 Block Diagram of the Project
FIGURE 5.16 Circuit Diagram of the Project
FIGURE 5.17 PDL of the Project
FIGURE 5.18 EasyPIC 5 Settings for the Button
FIGURE 5.19 Program Listing
FIGURE 5.20 Block Diagram of the Project
FIGURE 5.21 Circuit Diagram of the Project
FIGURE 5.22 PDL of the Project

FIGURE 5.23 Program Listing of the Project
FIGURE 5.24 EasyPIC 5 with LCD Display
FIGURE 5.25 Block Diagram of the Project
FIGURE 5.26 Sending Character "A" in Serial Format
FIGURE 5.27 RS232 Connectors
FIGURE 5.28 MAX232 Pin Configuration
FIGURE 5.29 Circuit Diagram of the Project
FIGURE 5.30 PDL of the Project
FIGURE 5.31 Program Listing
FIGURE 5.32 EasyPIC 5 Board Settings
FIGURE 5.33 The HyperTerminal
FIGURE 5.34 Select Serial Port
FIGURE 5.35 Set Serial Port Parameters
FIGURE 5.36 An Example Output from the Project

FIGURE 6.1 SIM Card layout
FIGURE 6.2 A Typical SIM Card Holder
FIGURE 6.3 The Smart GSM/GPRS Development Board
FIGURE 6.4 SIM340Z GSM/GPRS Modem Card
FIGURE 6.5 SIM340Z Functional Block Diagram
FIGURE 6.6 Smart GSM/GPRS Jumper and Switch Locations
FIGURE 6.7 Circuit Diagram of the Hardware Setup
FIGURE 6.8 Modem Responding With "OK"
FIGURE 6.9 Command ATI Displays Modem Product Information
FIGURE 6.10 Enabling and Disabling the Echo Mode
FIGURE 6.11 Displaying Current Value of Command Termination Character
FIGURE 6.12 Displaying Modem Manufacturer's Identification
FIGURE 6.13 Displaying the Modem Model Number
FIGURE 6.14 Sending an SMS Message
FIGURE 6.15 Listing Parameters of Command AT+CMGL
FIGURE 6.16 Listing all the Received SMS Messages on the Card
FIGURE 6.17 Display all SMS Messages
FIGURE 6.18 Display Message No 1
FIGURE 6.19 Display the SMS Service Center Number
FIGURE 6.20 Deleting Message No 1
FIGURE 6.21 Storing a Message on the SIM Card
FIGURE 6.22 Sending Message with id no 5 to a Mobile Phone
FIGURE 6.23 Sending a Message Using CMGW and CMSS
FIGURE 6.24 Format of a PDU Message
FIGURE 6.25 Sending SMS Message in PDU Mode
FIGURE 6.26 Circuit Diagram of the Project
FIGURE 6.27 Example Dialling a Mobile Phone
FIGURE 6.28 Displaying the Call Details

FIGURE 6.29 Command AT+CLTS
FIGURE 6.30 Command AT+CCVM
FIGURE 6.31 Command AT+CBAND
FIGURE 6.32 Command AT+CIMI
FIGURE 6.33 Command AT+CNUM
FIGURE 6.34 Command AT+CRSL
FIGURE 6.35 Command AT+CLVL
FIGURE 6.36 Command AT+CMUT

FIGURE 7.1 Block Diagram of the Project
FIGURE 7.2 Circuit Diagram of the Project
FIGURE 7.3 Connecting the EasyPIC 5 to the Smart GSM/GPRS Board
FIGURE 7.4 PDL of the Program
FIGURE 7.5 Program Listing of the Project
FIGURE 7.6 Block Diagram of the Project
FIGURE 7.7 LM35DZ Temperature Sensor
FIGURE 7.8 Circuit Diagram of the System
FIGURE 7.9 PDL of the Project
FIGURE 7.10 Program Listing of the Project
FIGURE 7.11 Message Sent to a Mobile Phone
FIGURE 7.12 Block Diagram of the Project
FIGURE 7.13 Circuit Diagram of the Project
FIGURE 7.14 PDL of the Project
FIGURE 7.15 Program Listing of the Project
FIGURE 7.16 Sending an SMS to a Mobile Phone

FIGURE 8.1 Pin Configuration of PCF8583
FIGURE 8.2 The Counter Registers
FIGURE 8.3 Format of the Hours Register
FIGURE 8.4 Format of the Year/Date Register
FIGURE 8.5 Format of the Weekdays/Month Register
FIGURE 8.6 Interfacing the RTC Chip to a PIC Microcontroller
FIGURE 8.7 Reading From the RTC Chip
FIGURE 8.8 Writing to the RTC Chip

CHAPTER 1

INTRODUCTION

1.1 Overview

A microcontroller is a single-chip computer. Before the invention of the microcontrollers most intelligent systems were designed based on microprocessors. A microprocessor is the processing element of a computer, consisting of an Arithmetic and Logic Unit (ALU) and the Control Unit (CU). A microprocessor on its own is useless and it has to be supported by a large number of chips, such as program and data memories, input-output circuits, clock circuit, interrupt circuits, timers etc.

A microcontroller is a computer in the form of a single chip and it includes all of the support and peripheral circuits required for a microprocessor. As a result, microcontrollers offer many advantages compared to microprocessors. Some of the advantages of microcontroller are: low-cost, low power consumption, simple design (only a few chips) and so on.

There are many microcontroller chips available from different manufacturers in the market. This book is about the use of PIC microcontrollers in GSM/GPRS modem based projects. The book describes the basic principles of PIC microcontrollers and shows with examples how these microcontrollers can be used in GSM/GPRS modem based projects. The main aim of the book is to show the reader how to design data logging systems using microcontrollers and GSM/GPRS modems. The remainder of this Chapter will describe the basic principles of data loggers to make the reader familiar with the concept.

The term *Data Logging* can be defined as the *capture and storage of data for use at a later time*. Basically, a data logger is an electronic device that captures and records data over time. Data loggers are nowadays based on the microcontroller technology. They are usually portable, battery operated devices with internal storage and some incorporating sensors to measure physical quantities such as temperature, pressure, humidity, flow, displacement, and so on.

In general, we can divide the data loggers into two basic groups:

1.2 Stand-alone Data Loggers

This type of data loggers can be used on their own without requiring other devices for data collection and storage. Stand-alone data loggers have large amounts of internal or external non-volatile memories (e.g. SD cards). They may also have real time clock chips. The collected data can be saved in the memory with time stamping.

The data collected in a stand-alone data logger is usually analyzed offline. A stand-alone data logger is usually configured and then left at the required site to collect data. At the end of the data collection period the device is connected to a PC and the collected data is read and analyzed offline using a program on the PC. Some stand-alone data loggers are dedicated for specific measurements, for example temperature data loggers. The Thermo Recorder TR-5 Series[1] is a typical stand-alone temperature data logger. This data logger has LCD output, it can record up to 16,000 readings with time intervals from 1 second to 1 hour and the battery life is quoted as 4 years.

One of the disadvantages of stand-alone data loggers is that the devices should be checked at regular intervals to make sure that the memory is not full, or the battery is not flat. This may sometimes cause problems since the device may be located at a remote location (such as at the top of a mountain, or at a place not easily accessible).

1.3 Data Capturing Data Loggers

Data capturing data loggers are used only to capture data. These devices do not have large internal memories and are normally connected to a PC for data storage. The data can either be analyzed offline or online. One of the disadvantages of data capturing data loggers is that the devices can not be used on their own as another device (e.g. a PC) is required to store the captured data. The Pico Technology DrDAQ[2] is a typical data capturing data logger which is connected to a PC to analyze the captured data. The device has built in sensors for light, sound, and temperature measurements.

Some data capturing data loggers have wireless capabilities. Usually a transmitter-receiver pair is used: the transmitter captures the data and sends it to the receiving device using wireless communication. The receiving device usually has large internal memory and stores the received data.

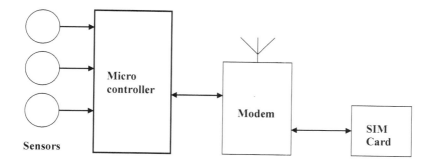

Figure 1.1 Block diagram of GSM/GPRS modem based data logger

Figure 1.1 shows the block diagram of a typical data logging system where data is read from the sensors and sent to a mobile phone in the form of SMS messages. In this book, a data logger system similar to the one given in Figure 1.1 will be developed. In addition, GSM/GPRS modem based projects will be given with complete hardware design and full program listings. The projects will enable the user to perform various activities using a GSM/GPRS modem. For example, calling a mobile phone, answering a phone call. Sending SMS messages. Reading SMS messages, and so on.

Chapter 2 provides a brief introduction to the microcontrollers where the basic building blocks of microcontrollers are described.

Chapter 3 is about the PIC series of microcontrollers and describes the basic features of the PIC16F887 microcontroller which is one of the low-cost and popular PIC16F model microcontroller. This microcontroller is used in all the projects in this book.

mikroC is one of the popular high level C language compilers for the PIC microcontrollers, developed by mikroElektronika (www.mikroe.com). Chapter 4 gives the basic principles of mikroC programming language with simple examples.

Chapter 5 is about the EasyPIC 5 microcontroller development board. This board is developed and manufactured by mikroElektronika and is one of the most popular microcontroller development boards available. The design of some simple projects using the EasyPIC 5 development board are also given in this Chapter.

Chapter 6 is about the Smart GSM/GPRS development board, developed and manufactured by mikroElektronika. This board enables a user to develop GSM and GPRS based projects in a relatively short time. In addition, various PC based projects using the Smart GSM/GPRS board are described in this Chapter.

Chapter 7 describes the design of projects using the Smart GSM/GPRS board and a microcontroller (e.g. the EasyPIC 5 development board).

Chapter 8 is about the PCF8583 real-time clock chip (RTC). The Chapter describes the basic features of this chip and explains how the chip can be used to provide RTC clock data to a microcontroller based system.

CHAPTER 2

MICROCONTROLLERS

2.1 Overview

The term microcontroller is used to describe a system that includes a minimum of a microprocessor, program memory, data memory, and input-output (I/O). Some microcontroller systems include additional components such as timers, counters, analogue-to-digital converters, and so on. Thus, a microcontroller system can be anything from a large computer having hard disks, floppy disks, and printers, to a single chip embedded controller.

In this book we are going to consider only the type of microcontroller that consists of a single silicon chip. Such microcontroller systems are used in many household goods such as microwave ovens, TV remote control units, cookers, hi-fi equipment, CD players, personal computers, fridges, etc. There are a large number of microcontrollers available in the market. In this book we shall be looking at the programming and system design using the PIC16F (*programmable interface controller*) series of microcontrollers (the popular PIC16F887) manufactured by *Microchip Technology Inc*[3]. (www.microchip.com)

2.2 Microcontroller Systems

A microcontroller is a single chip computer. *Micro* suggests that the device is small, and *controller* suggests that the device can be used in control applications. Another term used for microcontrollers is *embedded controller*, since most of the microcontrollers are built into (or embedded in) the devices they control.

A microprocessor differs from a microcontroller in many ways. The main difference is that a microprocessor requires several other components for its operation, such as program memory and data memory, input-output devices, and external clock circuit. A microcontroller on the other hand has all the support chips incorporated inside the same chip. All microcontrollers operate on a set of instructions (or the user program) stored in their memory. A microcontroller fetches the instructions from its program memory one by one, decodes these instructions, and then carries out the required operations.

Microcontrollers have traditionally been programmed using the assembly language of the target device. Although the assembly language is fast, it has several disadvantages. An assembly program consists of mnemonics and it is difficult to learn and maintain a program written using the assembly language. Also, microcontrollers manufactured by different firms have different assembly languages and the user is required to learn a new language every time a new microcontroller is to be used. Microcontrollers can also be programmed using a high-level language, such as BASIC, PASCAL, and C. High-level languages have the advantage that it is much easier to learn a high-level language than an assembler. Also, very large and complex programs can easily be developed using a high-level language. In this book we shall be learning the programming of PIC microcontrollers using the popular **mikroC** C language, developed by **mikroElektronika**[4].

In general, a single chip is all that is required to have a running microcontroller system. In practical applications additional components may be required to allow a microcontroller to interface to its environment. With the advent of the PIC family of microcontrollers the development time of an electronic project has reduced to several hours.

Basically, a microcontroller executes a user program which is loaded in its program memory. Under the control of this program data is received from external devices (inputs), manipulated and then sent to external devices (outputs).

The simplest microcontroller architecture consists of a microprocessor, memory, and input-output. The microprocessor consists of a central processing unit (CPU), and the control unit (CU). The CPU is the brain of the microcontroller and this is where all of the arithmetic and logic operations are performed. The control unit controls the internal operations of the microprocessor and sends out control signals to other parts of the microcontroller to carry out the required instructions

Memory is an important part of a microcontroller system. Depending upon the type used, we can classify memories into two groups: program memory, and data memory. Program memory stores the program written by the programmer and this memory is usually non-volatile. i.e. data is not lost after the removal of power. Data memory is where the temporary data used in a program are stored and this memory is usually volatile. i.e. data is lost after the removal of power.

There are basically six types of memories as summarised below.

2.2.1 RAM

RAM means Random Access Memory. It is a general purpose memory which usually stores the user data in a program. RAM memory is volatile in the sense that it cannot retain data in the absence of power. i.e. data is lost after the removal of power. Most microcontrollers have some amount of internal RAM. 256 bytes is a common amount, although some microcontrollers have more, some less. For example, the PIC18F452 microcontroller has 1536 bytes of RAM. In general it is possible to extend the memory by adding external memory chips.

2.2.2 ROM

ROM is Read Only Memory. This type of memory usually holds program or fixed user data. ROM is non-volatile. If power is removed from ROM and then reapplied, the original data will still be there. ROM memories are programmed at factory during the manufacturing process and their contents can not be changed by the user. ROM memories are only useful if you have developed a program and wish to order several thousand copies of it.

2.2.3 PROM

PROM is programmable Read Only Memory. This is a type of ROM that can be programmed in the field, often by the end user, using a device called a PROM programmer. Once a PROM has been programmed, its contents cannot be changed. PROMs are usually used in low production applications where only several such memories are required.

2.2.4 EPROM

EPROM is Erasable Programmable Read Only Memory. This is similar to ROM, but the EPROM can be programmed using a suitable programming device. EPROM memories have a small clear glass window on top of the chip where the data can be erased under strong ultraviolet light. Once the memory is programmed, the window can be covered with dark tape to prevent accidental erasure of the data. An EPROM memory must be erased before it can be re-programmed. Many development versions of microcontrollers are manufactured with EPROM memories where the user program can be stored. These memories are erased and re-programmed until the user is satisfied with the program. Some versions of EPROMs, known as OTP (One Time Programmable), can be programmed using a suitable programmer device but these memories can not be erased. OTP memories cost much less than the EPROMs. OTP is useful after a project has been developed completely and it is required to make many copies of the program memory.

2.2.5 EEPROM

EEPROM is Electrically Erasable Programmable Read Only Memory, which is a non-volatile memory. These memories can be erased and also be re-programmed using suitable programming devices. EEPROMs are used to save configuration information, maximum and minimum values, identification data etc. Some microcontrollers have built-in EEPROM memories. e.g. PIC18F452 contains a 256-byte EEPROM memory where each byte can be programmed and erased directly by applications software. EEPROM memories are usually very slow. The cost of an EEPROM chip is much higher than that of an EPROM chip.

2.2.6 Flash EEPROM

This is another version of EEPROM type memory. This memory has become popular in microcontroller applications and is used to store the user program. Flash EEPROM is non-volatile and is usually very fast. The data can be erased and then re-programmed using a suitable programming device. Some microcontrollers have only 1K flash EEPROM while some others have 32K or more. PIC18F452 microcontroller has 32K bytes of flash memory.

2.3 Microcontroller Features

Microcontrollers from different manufacturers have different architectures and different capabilities. Some may suit a particular application while others may be totally unsuitable for the same application. The hardware features of microcontrollers in general are described in this section.

2.3.1 Supply Voltage

Most microcontrollers operate with the standard logic voltage of +5V. Some microcontrollers can operate at as low as +2.7V and some will tolerate +6V without any problems. You should check the manufacturers' data sheets about the allowed limits of the power supply voltage. For example, PIC18F452 microcontrollers can operate with a power supply +2V to +5.5V.

A voltage regulator circuit is usually used to obtain the required power supply voltage when the device is to be operated from a mains adaptor or batteries. For example, a 5V regulator (e.g. an LM7805, or the low power version LM78L05, or a similar regulator) is required if the microcontroller is to be operated from a 5V supply using a 9V battery.

2.3.2 The Clock

All microcontrollers require a clock (or an oscillator) to operate. The clock is usually provided by connecting external timing devices to the microcontroller. Most microcontrollers will generate clock signals when a crystal and two small capacitors are connected. Some will operate with resonators or external resistor-capacitor pair. Some microcontrollers have built-in timing circuits and they do not require any external timing components. If your application is not time sensitive you should use external or internal (if available) resistor-capacitor timing components for simplicity and low cost.

An instruction is executed by fetching it from the memory and then decoding it. This usually takes several clock cycles and is known as the *instruction cycle*. In PIC microcontrollers an instruction cycle takes four clock periods. Thus, the microcontroller is actually operated at a clock rate which is a quarter of the actual oscillator frequency. PIC16 series of microcontrollers can usually operate with clock frequencies up to 20MHz. Higher end PIC microcontrollers, such as the PIC18F series can operate at frequencies up to 40MHz.

2.3.3 Timers

Timers are important parts of any microcontroller. A timer is basically a counter which is driven either from an external clock pulse or from the internal oscillator of the microcontroller. A timer can be 8-bits, or 16-bits wide. Data can be loaded into a timer under program control and the timer can be stopped or started by program control. Most timers can be configured to generate an interrupt when they reach a certain count (usually when they overflow). The interrupt can be used by the user program to carry out accurate timing related operations inside the microcontroller. PIC16F and PIC18F series of microcontrollers usually have 3 timers. For example, PIC16F877 has two 8-bit and one 16-bit timers. Similarly, PIC18F452 microcontroller has 3 built-in timers.

Some microcontrollers offer capture and compare facilities where a timer value can be read when an external event occurs, or the timer value can be compared to a preset value and an interrupt can be generated when this value is reached. Most PIC16 and PIC18 series of microcontrollers have at least 2 capture and compare modules.

2.3.4 Watchdog

Most microcontrollers have at least one watchdog facility. The watchdog is basically a timer which is refreshed by the user program and a reset occurs if the program fails to refresh the watchdog. The watchdog timer is used to detect a system problem, such as the program being in an endless loop. A watchdog is a safety feature that prevents runaway software and stops the microcontroller from executing meaningless and unwanted code. Watchdog facilities are commonly used in real-time systems where it is required to regularly check the successful termination of one or more activities.

2.3.5 Reset Input

A reset input is used to reset a microcontroller externally. Resetting puts the microcontroller into a known state such that the program execution starts from address 0 of the program memory. An external reset action is usually achieved by connecting a push-button switch to the reset input such that the microcontroller can be reset when the switch is pressed.

2.3.6 Interrupts

Interrupts are very important concepts in microcontrollers. An interrupt causes the microcontroller to respond to external and internal (e.g. a timer) events very quickly. When an interrupt occurs the microcontroller leaves its normal flow of program execution and jumps to a special part of the program, known as the *Interrupt Service Routine* (ISR). The program code inside the ISR is executed and upon return from the ISR the program resumes its normal flow of execution.

The ISR starts from a fixed address of the program memory. This address is also known as the *interrupt vector address*. Some microcontrollers with multi-interrupt features have just one interrupt vector address, while some others have unique interrupt vector addresses, one for each interrupt source. Interrupts can be nested such that a new interrupt can suspend the execution of another interrupt. Another important feature of a microcontroller with multi-interrupt capability is that different interrupt sources can be given different levels of priority. For example, PIC18F series of microcontrollers have low-priority and high-priority interrupt levels.

2.3.7 Brown-out Detector

Brown-out detectors are also common in many microcontrollers and they reset a microcontroller if the supply voltage falls below a nominal value. Brown-out detectors are safety features and they can be employed to prevent unpredictable operation at low voltages, especially to protect the contents of EEPROM type memories.

2.3.8 Analogue-to-digital Converter

An analogue-to-digital converter (A/D) is used to convert an analogue signal such as voltage to a digital form so that it can be read and processed by a microcontroller. Some microcontrollers have built-in A/D converters. It is also possible to connect an external A/D converter to any type of microcontroller. A/D converters are usually 8 to 10 bits, having 256 to 1024 quantisation levels. Most PIC microcontrollers with A/D features have multiplexed A/D converters where more than one analogue input channel is provided. For example, PIC16F877 microcontroller has 14 10-bit A/D converter channels. Similarly, PIC18F452 microcontroller has 8 10-bit A/D converter channels.

The A/D conversion process must be started by the user program and it may take several hundreds of microseconds for a conversion to complete. A/D converters usually generate interrupts when a conversion is complete so that the user program can read the converted data quickly.

A/D converters are very useful in control and monitoring applications since most sensors (e.g. temperature sensor, pressure sensor, force sensor etc.) produce analogue output voltages.

2.3.9 Serial Input-Output

Serial communication (also called RS232 communication) enables a microcontroller to be connected to another microcontroller or to a PC using a serial cable. Some microcontrollers have built-in hardware called USART (Universal Synchronous-Asynchronous Receiver-Transmitter) to implement a serial communication interface. The baud rate and the data format can usually be selected by the user program. If any serial input-output hardware is not provided, it is easy to develop software to implement serial data communication using any I/O pin of a microcontroller. Most PIC16F and PIC18F series of microcontrollers have built-in USART modules.

Some microcontrollers (e.g. PIC18F series) incorporate SPI (Serial Peripheral Interface) or I^2C (Integrated Inter Connect) hardware bus interfaces. These enable a microcontroller to interface to other compatible devices easily.

2.3.10 EEPROM Data Memory

EEPROM type data memory is also very common in many microcontrollers. The advantage of an EEPROM memory is that the programmer can store non-volatile data in such a memory, and can also change this data whenever required. For example, in a temperature monitoring application the maximum and the minimum temperature readings can be stored in an EEPROM memory. Then, if the power supply is removed for whatever reason, the values of the latest readings will still be available in the EEPROM memory. PIC16F877 microcontroller has 256 bytes of EEPROM memory. Similarly, PIC18F452 microcontroller has 256 bytes of EEPROM memory. Some other members of the family have more (e.g. PIC18F6680 has 1024 bytes) EEPROM memories.

mikroC language provides special instructions for reading and writing to the EEPROM memory of a PIC microcontroller.

2.3.11 LCD Drivers

LCD drivers enable a microcontroller to be connected to an external LCD display directly. These drivers are not common since most of the functions provided by them can be implemented in software. For example, PIC18F6490 microcontroller has built-in LCD driver module.

2.3.12 Analogue Comparator

Analogue comparators are used where it is required to compare two analogue voltages. Although these circuits are implemented in most high-end PIC microcontrollers they are not common in other microcontrollers. Most PIC16F and PIC18F series of microcontrollers have built-in analog comparator modules.

2.3.13 Real-time Clock

Real-time clock enables a microcontroller to have absolute date and time information continuously. Built-in real-time clocks are not common in most microcontrollers since they can easily be implemented by either using a dedicated real-time clock chip, or by writing a program.

2.3.14 Sleep Mode

Some microcontrollers (e.g. PIC) offer built-in sleep modes where executing this instruction puts the microcontroller into a mode where the internal oscillator is stopped and the power consumption is reduced to an extremely low level. The main reason of using the sleep mode is to conserve the battery power when the microcontroller is not doing anything useful. The microcontroller usually wakes up from the sleep mode by external reset or by a watchdog time-out.

2.3.15 Power-on Reset

Some microcontrollers (e.g. PIC) have built-in power-on reset circuits which keep the microcontroller in reset state until all the internal circuitry has been initialised. This feature is very useful as it starts the microcontroller from a known state on power-up. An external reset can also be provided where the microcontroller can be reset when an external button is pressed.

2.3.16 Low Power Operation

Low power operation is especially important in portable applications where the microcontroller based equipment is operated from batteries. Some microcontrollers (e.g. PIC) can operate with less than 2mA with 5V supply, and around 15µA at 3V supply. Some other microcontrollers, especially microprocessor based systems where there could be several chips may consume several hundred milliamperes or even more.

2.3.17 Current Sink/Source Capability

This is important if the microcontroller is to be connected to an external device which may draw large current for its operation. PIC microcontrollers can source and sink 25mA of current from each output port pin. This current is usually sufficient to drive LEDs, small lamps, buzzers, small relays etc. The current capability can be increased by connecting external transistor switching circuits or relays to the output port pins.

2.3.18 USB interface

USB is currently a very popular computer interface specification used to connect various peripheral devices to computers and microcontrollers. Some PIC microcontrollers provide built-in USB modules. For example, PIC18F2x50 has built-in USB interface capabilities.

2.3.19 Motor control interface

Some PIC microcontrollers, for example PIC18F2x31 provide motor control interface.

2.3.20 CAN interface

CAN bus is a very popular bus system used mainly in automation applications. Some PIC18F series of microcontrollers (e.g. PIC18F4680) provide CAN interface capabilities.

2.3.21 Ethernet interface

Some PIC microcontrollers (e.g. PIC18F97J60) provide Ethernet interface capabilities. Such microcontrollers can easily be used in network based applications.

2.3.22 ZigBee interface

ZigBee is an interface similar to Bluetooth and is used in low-cost wireless home automation applications. Some PIC18F series of microcontrollers provide ZigBee interface capabilities making the design of such wireless systems very easy.

2.3.23 Pulse-Width-Modulated (PWM) output

Some PIC16F series and all PIC18F series microcontrollers provide PWM outputs where the frequency and the duty cycle of the generated waveform can be programmed.

2.3.24 In-Circuit Serial Programming (ICSP)

In-Circuit Serial Programming (ICSP) is a technique used to program a microcontroller while the device is not removed from the applications circuit. ICSP uses a few microcontroller I/O pins to program the device. Most PIC16F series and all PIC18F series of microcontrollers support ICSP.

2.4 Microcontroller Architectures

Usually two types of architectures are used in microcontrollers (see Figure 2.1): *Von Neumann* architecture and *Harvard* architecture. Von Neumann architecture is used by a large percentage of microcontrollers and here all memory space is on the same bus and instruction and data use the same bus. In the Harvard architecture (used by the PIC microcontrollers), code and data are on separate busses and this allows the code and data to be fetched simultaneously, resulting in an improved performance.

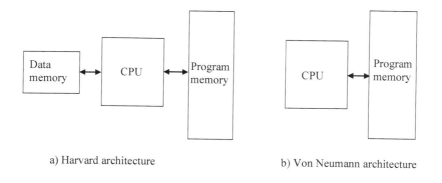

a) Harvard architecture   b) Von Neumann architecture

Figure 2.1 Von Neumann and Harvard architectures

2.4.1 RISC and CISC

RISC (Reduced Instruction Set Computer) and CISC (Complex Instruction Computer) refer to the instruction set of a microcontroller. In an 8-bit RISC microcontroller, data is 8-bits wide but the instruction words are more than 8-bits wide (usually 12, 14 or 16-bits) and the instructions occupy one word in the program memory. Thus, the instructions are fetched and executed in one cycle, resulting in an improved performance.

In a CISC microcontroller both data and instructions are 8-bits wide. CISC microcontrollers usually have over 200 instructions. Data and code are on the same bus and can not be fetched simultaneously.

CHAPTER 3

THE PIC16F887 MICROCONTROLLER

3.1 Overview

PIC16F887 microcontroller is a member of the family PIC16F882/883/884/886/887. The family consists of CMOS 8-bit microcontrollers with 28, 40 and 44 pins. All the members of the family have similar architectures with different capacities. Table 3.1 summarizes the main differences between various members of this family.

Table 3.1 PIC16F887 family of microcontrollers

Device	Program memory (Flash)	Data memory (RAM)	Data memory (EEPROM)	I/O	A/D	USART
PIC16F882	2048	128	128	24	11	1
PIC16F883	4096	256	256	24	11	1
PIC16F884	4096	256	256	35	14	1
PIC16F886	9192	368	256	24	11	1
PIC16F887	8192	368	256	35	14	1

PIC16F887 is a 40-pin highly popular microcontroller used in medium speed and medium complexity applications. In this book, the PIC16F887 microcontroller is used in all of the projects. The architecture and the various features of this microcontroller are given in detail in the following sections. Further information about the PIC16F887 family[3] can be obtained from the Microchip Document DS41291F, entitled *"PIC16F882/883/884/886/887 Data Sheet, 2009"*.

3.2 The Specifications of the PIC16F887 Microcontroller

The PIC16F887 microcontroller has the following basic features:

- 40-pins
- RISC instruction set (only 35 instructions)
- 8192 words program memory (flash)
- 368 bytes RAM memory
- 256 bytes EEPROM memory

- 35 I/O pins
- 14 10-bit A/D converters
- 2 analog comparators
- Capture, compare, PWM modules
- USART module
- Up to 20MHz operation
- Internal clock (up to 8MHz)
- Power-on-reset (POR)
- Power-up-timer (PWRT)
- Watchdog timer
- Low voltage (2V) operation
- In-Circuit Serial Programming
- High I/O source and sink currents

3.3 The Architecture of the PIC16F887 Microcontroller

Figure 3.1 shows the internal architecture of the PIC16F887 microcontroller. The program counter is 13-bits wide, capable of accessing 8k of program memory (0000 to 1FFFh). The Reset vector is located at address 0 of the program memory where the first executable instruction is locate. The interrupt vector is at address 0004h of the program memory.

The data memory consists of 4-banks, each bank 128 bytes long. The initial part of each bank is reserved and used by the Special Function Registers (SFR), leaving a total of 368 bytes for user data area. The bank selection is controlled by the STATUS register bits RP0 and RP1. The SFR registers are used by the CPU and peripheral devices, such as I/O control, USART, A/D control and so on.

Some important SFR registers are the STATUS register, OPTION register, and the INTCON register. The STATUS register controls the data memory bank selection, power-down control, zero status bit, and the carry-borrow bit. The OPTION register controls the PORT B pull-ups, external interrupt edge select control, Timer 0 clock source and clock edge select control, and Timer 0 Prescaler rate select bits. The INTCON register is used to enable and disable the various interrupt sources in addition to enabling or disabling the global interrupt requests to the CPU.

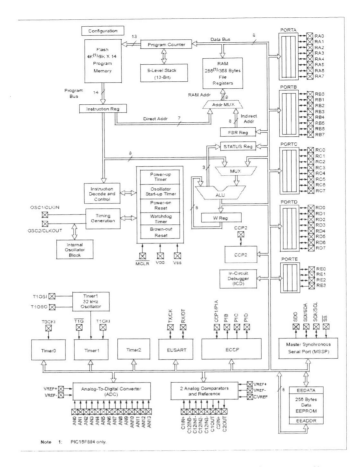

Figure 3.1 Architecture of the PIC16F887 microcontroller
(www.microchip.com)

The PIC16F887 microcontroller has 35 digital I/O bits, organized as 8-bit and 3-bit ports as follows:

PORT A	-	8 bits
PORT B	-	8 bits
PORT C	-	8 bits
PORT D	-	8 bits
PORT E	-	3 bits

Most of the I/O bits have multiple functions and the function of an I/O bit can be selected by the appropriate SFR register bits. For example, analog ports AN0 – AN4 share the four low bits of PORT A. AN5 – AN7 share the PORT E bits, and AN8 – AN13 share the PORT B bits. Similarly, PORT C bit RC6 is shared by the USART transmit data bit TXD, and PORT C bit RC7 is shared by the USART receive data bit RXD.

On power-up, or after a system Reset, the ports with analog functionalities are automatically configured as analog ports. In order to configure these ports as digital we have to program the appropriate SFR register bits. For example, SFR register ANSEL is used to configure analog ports AN0 – AN7. Clearing a bit of ANSEL to 0 configures the corresponding port pin to be digital I/O, and setting a bit of ANSEL to 1 configures the corresponding port pin to be analog. Similarly, analog I/O pins AN8 – AN13 are configured using the SFR register ANSELH.

The SFR registers TRISx are used to configure the direction of a port pin. Every port has an associated TRIS register. Thus the existing TRIS registers are: TRISA, TRISB, TRISC, TRISD, and TRISE. Setting a bit in a TRISx register to a 1 configures the direction of the corresponding port pin to be an input pin. Similarly, clearing a bit in a TRISx register to 0 configures the direction of the corresponding port pin to be an output pin. For example, the instruction TRISB = 0xFF sets all bits of TRISB to 1, thus making all PORT B I/O pins to be input pins. Similarly, the instruction TRISB = 0xF0 sets all high 4 bits of TRISB to 1 and clears the low 4 bits to 0, thus making the upper 4 PORT B pins to be input pins, and the lower 4 PORT B pins to be output pins.

3.4 Clock Configurations

The PIC16F887 microcontroller can be driven in various clock configurations as listed below:

- External clock source (EC mode)
- 32kHz Low Power crystal (LP mode)
- Medium speed crystal or resonator (XT mode)
- High speed crystal or resonator (HS mode)
- Resistor-capacitor (RC mode)
- Resistor-capacitor with I/O (RCIO mode)
- Internal oscillator (INTOSC mode)
- Internal oscillator with I/O (INTOSCIO mode)

Oscillator modes are programmed by the SFR register OSCCON as follows:

Bit 7   Unused

Bit 6-4   111 = 8MHz
110 = 4MHz
101 = 2 MHz
100 = 1 MHz
011 = 500 kHz
010 = 250 kHz
001 = 125 kHz
000 = 31 kHz

Bit 3   1 = device is running from external clock (read-only)
0 = device is running from internal clock (read-only)

Bit 2   1 = Internal oscillator is stable (read-only)
0 = Internal oscillator is not stable (read-only)

Bit 1   1 = Low frequency (31kHz) internal oscillator is stable (read-only)
0 = Low frequency (31 kHz) internal oscillator is not stable (read-only)

Bit 0   1 = Select internal oscillator as clock source
0 = Select external oscillator as clock source

By default the external oscillator is selected after a power-up and CPU register CONFIG1 determines the type of external clock used (e.g. EC mode, crystal, RC mode etc).

The device can be operated with a crystal at up to 20MHz. In low speed non-critical operations the clock pulses can be supplied by using an external R-C circuit connected to pin OSC1 of the microcontroller. The clock frequency is then dependent on the external resistor and capacitor used, the temperature, and the supply voltage. The value of the external resistor should be between 3K and 100K, and the capacitor should be greater than 20pF. Figure 3.2 shows the typical configuration using the R-C circuit. In this configuration, pin OSC2 of the microcontroller provides a clock output at a rate of f/4. As shown in Figure 3.3, the microcontroller can also be operated in what is knows as the RCIO clock mode where pin OSC2 becomes a general purpose I/O pin.

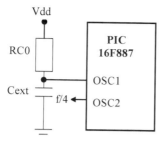

Figure 3.2 R-C clock mode

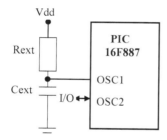

Figure 3.3 RCIO clock mode

In frequencies at up to 4MHz, a resonator or a crystal can be used to provide the timing pulses (XT mode). Figure 3.4 shows how a resonator can be connected to the microcontroller.

Figure 3.4 Using a resonator

In high frequency and high precision applications (XT mode or HS mode), a crystal should be connected between pins OSC1 and OSC2 with series capacitors as shown in Figure 3.5. The value of the capacitors should be around 15 - 22pF each.

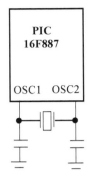

Figure 3.5 Using a crystal

The PIC16F887 microcontroller can also be operated with externally supplied clock pulses (EC mode) as shown in Figure 3.6.

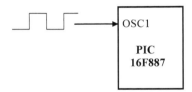

Figure 3.6 Using external clock

It is also possible to operate the PIC16F877 microcontroller using its built-in internal oscillator. The built-in oscillator supports operations from 31 kHz to 8 MHz (the default power-on value is 4 MHz). Although the internal high frequency clock is factory calibrated, it can be adjusted using the SFR register OSCTUNE. The system clock can be switched between external and internal clock sources via software using bit 0 of SFR register OSCCON. More information on the use of the internal clock sources can be obtained from the manufacturers' data sheet.

3.5 Reset Configurations

The PIC18F4520 microcontroller is reset by one of the following conditions:

- MCLR reset
- Power-on reset

- Watchdog reset
- Brown-out reset
- Stack full reset
- Stack underflow reset
- Reset instruction

In this section we shall be looking at how the microcontroller can be reset at power-up and also by using an external reset button.

3.5.1 Power-on Reset

The power-on reset takes place when power is applied to the microcontroller. To use the power-on reset feature of the microcontroller, the MCLR input pin should be connected to the supply voltage through a 1K to 10K resistor. Figure 3.7 shows a typical power-on reset circuit.

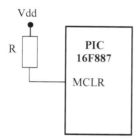

Figure 3.7 Power-on reset

In some applications where the power supply rise time is slow, it may be required to use a diode, resistors, and a capacitor as shown in Figure 3.8.

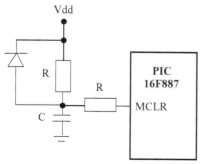

Figure 3.8 Power-on reset for slow rising Vdd

3.5.2 External Reset

In many applications it may be required to reset the microcontroller externally by pressing a button. Figure 3.9 shows how the external reset action can be done using a push-button switch. In this circuit, the MCLR input of the microcontroller is normally at logic 1. Pressing the RESET button forces the MCLR button to logic 0 which resets the microcontroller. Leaving the button resumes operation where execution starts from address 0 of the program memory.

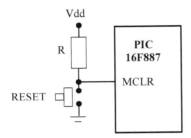

Figure 3.9 External reset circuit

3.6 Minimum Crystal Based Configuration

Figure 3.10 shows the minimum crystal based microcontroller configuration with power-on reset circuit, operating at 4MHz.

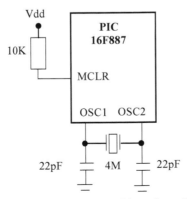

Figure 3.10 Minimum crystal based configuration

CHAPTER 4

mikroC LANGUAGE FOR PIC16 MICROCONTROLLERS

4.1 Overview

There are several C compilers in the market for the PIC16 series of microcontrollers. Most of the features of these compilers are similar and they can all be used to develop C based high-level programs for the PIC16 series of microcontrollers.

Some of the popular C compilers used in the development of commercial, industrial, and educational PIC16 microcontroller applications are (these compilers can be used for both PIC16 and PIC18 devices):

- mikroC C compiler
- PICC18 C compiler
- C18 C compiler
- CCS C compiler

mikroC C compiler[4] has been developed by *mikroElektronika* (web site: www.microe.com) and is one of the easy to learn compilers with rich resources, such as a large number of library functions and an integrated development environment with built-in simulator, and an in-circuit-debugger (e.g. mikroICD). A demo version of the compiler with a 2K program limit is available from mikroElektronika. In this book we shall be mainly concentrating on the use of the microC compiler and all of the projects described in the book are based on this compiler.

PICC18 C compiler is another popular C compiler, developed by *Hi-Tech Software*[5] (web site: www.htsoft.com). This compiler has two versions: the standard compiler, and the professional version. A powerful simulator and an integrated development environment (Hi-Tide) is provided by the company. PICC18 is supported by the PROTEUS simulator (www.labcenter.co.uk) that can be used to simulate PIC microcontroller based systems. A limited period demo version of this compiler is available from the developers' web site.

C18 C compiler[3] is a product of the *Microchip Inc.* (web site: www.microchip.com). A limited period demo version, and a limited functionality version with no time limit of C18 are available from the Microchip web site.

CCS C compiler[6] has been developed by the *Custom Computer Systems Inc* (web site: www.ccsinfo.com). The company provides a limited period demo version of their compiler. CCS compiler provides a large number of build-in functions and supports an in-circuit-debugger (e.g. ICD-U40) which aids greatly in the development of PIC16 and PIC18 microcontroller based systems.

4.2 Structure of a mikroC Program

Figure 4.1 shows the simplest structure of a mikroC program. This program flashes an LED connected to port RC0 (bit 0 of PORT C) of a PIC microcontroller with one second intervals. Do not worry if you do not understand the operation of the program at this stage as all will be clear as we progress through this chapter. Some of the programming elements used in Figure 4.1 are described below in detail.

```
/**********************************************
              LED FLASHING PROGRAM
              ************************

This program flashes an LED connected to port pin RC0 of PORT C
with one second intervals.

    Programmer    : D. Ibrahim
    File          : LED.C
    Micro         : PIC16F887
***********************************************/

void main()
{
   for(;;)                      // Endless loop
   {
       TRISC = 0;               // Configure PORT C as output
       PORTC.0 = 0;             // RC0 = 0
       Delay_Ms(1000);          // Wait 1 second
       PORTC.0 = 1;             // RC0 = 1
       Delay_Ms(1000);          // Wait 1 second
   }                            // End of loop
}
```

Figure 4.1 Structure of a simple mikroC program

At the beginning of the program PORT C pins are configured as outputs with the statement TRISC = 0. Then bit 0 of PORT C is cleared which turns OFF the LED. After a 1 second delay with the Delay_Ms(1000) statement, bit 0 of PORT C is set to logic 1 which turns ON the LED. This process is repeated after one second delay.

4.2.1 Comments

Comments are used by programmers to clarify the operation of the program or a programming statement. Comment lines are ignored and not compiled by the compiler. Two types of comments can be used in mikroC programs: long comments extending several lines, and short comments occupying only a single line. Comment lines are usually used at the beginning of a program to describe briefly the operation of the program, the name of the author, the program filename, the date program was written, and a list of version numbers together with the modifications in each version. As shown in Figure 4.1, comments can also be used after statements to describe the operations performed by the statements. A well commented program is important for the maintenance and thus for the future lifetime of a program. In general, any programmer will find it easier to modify and/or update a well commented program.

As shown in Figure 4.1, long comments start with characters "/*" and terminate with characters "*/". Similarly, short comments start with characters "//" and there is no need to terminate short comments.

4.2.2 Beginning and Ending of a Program

In C language a program begins with the keywords:

    void main()

After this, a curly opening bracket is used to indicate the beginning of the program body. The program is terminated with a closing curly bracket. Thus, as shown in Figure 4.1, the program has the following structure:

    void main()
    {
        program body
    }

4.2.3 Terminating Program Statements

In C language all program statements must be terminated with the semicolon (";") character, otherwise a compiler error will be generated:

```
j = 5;           // correct
j = 5            // error
```

4.2.4 White Spaces

White spaces are spaces, blanks, tabs, and newline characters. All white spaces are ignored by the C compiler. Thus, the following three sequences are all identical:

```
int i;      char j;
```

or,

```
int i;
char j;
```

or,

```
int i;
        char j;
```

Similarly, the following sequences are identical:

```
i = j + 2;
```

or,

```
i = j
    + 2;
```

4.2.5 Case Sensitivity

In general C language is case sensitive and variables with lower case names are different from those with upper case names. Currently mikroC variables are not case sensitive and the following variables are all equivalent. It is believed that future releases of mikroC will offer case sensitivity:

```
total    TOTAL    Total    ToTal    total    totaL
```

The only exception is that identifiers **main** and **interrupt** must be written in lower case in mikroC. In this book we shall be assuming that the variables are case sensitive for compatibility with other C compilers, and variables with same names but different cases shall not be used.

4.2.6 Variable Names

In C language variable names can begin with an alphabetical character or with the underscore character. In essence, variable names can be any of the characters a-z and A-Z, the digits 0-9 and the underscore character " _ ". Each variable name should be unique within the first 31 characters of its name. Variable names can contain upper case and lower case characters (see 4.1.5 above) and numeric characters can be used inside a variable name. Examples of valid variable names are:

    Sum   count   sum100   counter   i1   UserName   _myName

Examples of invalid variable names are:

    1Book   @count   ?sum   235total $name   ~abc

Some names are reserved for the compiler itself and they can not be used as variable names in our programs. Table 4.1 gives a list of these reserved words.

Table 4.1 mikroC reserved names

asm	enum	signed
auto	extern	sizeof
break	float	static
case	for	struct
char	goto	switch
const	if	typedef
continue	int	union
default	long	unsigned
do	register	void
double	return	volatile
else	short	while

As an example, the following statements are not valid:

    goto = 5

or,

    switch = if + 1

4.2.7 Variable Types

mikroC language supports the variable types shown in Table 4.2. Details and examples of these variables are given in this section.

Table 4.2 mikroC variable types

Type	Size (bits)	Range
unsigned char	8	0 to 255
unsigned short int	8	0 to 255
unsigned int	16	0 to 65535
unsigned long int	32	0 to 4294967295
signed char	8	-128 to 127
signed short int	8	-128 to 127
signed int	16	-32768 to 32767
signed long int	32	-2147483648 to 2147483647
Float	32	±1.17549435082E-38 to ±6.80564774407E38
Double	32	±1.17549435082E-38 to ±6.80564774407E38
long double	32	±1.17549435082E-38 to ±6.80564774407E38

(unsigned) char or unsigned short (int)

These are 8-bit unsigned variables with a range 0 to 255. In the following example two 8-bit variables named **total** and **sum** are created and **sum** is assigned decimal value 150:

    unsigned char total, sum;
    sum = 150;
or,
    char total, sum;
    sum = 150;

Variables can be assigned values during their declaration. Thus, the above statements can also be written as:

    char total, sum = 150;
or,
    unsigned char sum = 150;

signed char or (signed) short (int)

These are 8-bit signed character variables with a range -128 to +127. In the following example a signed 8-bit variable named **counter** is created with a value of -50:

      signed char counter = -50;

or,

      short counter = -50;

or,

      short int counter = -50;

(signed) int

These are 16-bit variables with a range -32768 to +32767. In the following example a signed integer named **Big** is created:

      int Big;

unsigned (int)

These variables are unsigned 16-bit with a range 0 to 65535. In the following example an unsigned 16-bit variable named **count** is created and is assigned value 12000:

      unsigned int count = 12000;

(signed) long (int)

These variables are 32-bits long with a range -2147483648 to + 2147483647. An example is given below:

      signed long LargeNumber;

unsigned long (int)

These are 32-bit unsigned variables having the range 0 to 4294967295. An example is given below:

      unsigned long VeryLargeNumber;

or,

      unsigned long VeryLargeNumber = 123456789;

float or **double** or **long double**

These are floating point variables, implemented in mikroC using Microchip AN575 32-bit format which is IEEE 754 compliant. Floating point numbers range from ±1.17549435082E-38 to ±6.80564774407E38. In the following example a floating point variable named **area** is created and is assigned value 12.235:

> **float** area;
> area = 12.235;

In order to avoid confusion during program development it is recommended that you specify the sign of a variable type (signed or unsigned), followed by the type of the variable. For example, use **unsigned char** instead of **char** only. Similarly, use **unsigned int** instead of **unsigned** only.

In this book we shall be using the following mikroC data types which are easier to remember, and are also compatible with most other C compilers:

unsigned char	0 to 255
signed char	-128 to 127
unsigned int	0 to 65535
signed int	-32768 to 32767
unsigned long	0 to 4294967295
signed long	-2147483648 to 2147483647
float	±1.17549435082E-38 to ±6.80564774407E38

4.2.8 Constants

Constants represent fixed values (numeric or character) in programs that can not be changed. Constants are stored in the flash program memory of the PIC microcontroller, thus the valuable and limited RAM memory is not wasted. In mikroC constants can be: integer, floating point, character, string, or enumerated types.

Integer Constants

Integer constants can be decimal, hexadecimal, octal, or binary. The data type of a constant is derived by the compiler from its value. But, suffixes can be used to change the type of a constant.

From Table 3.2 we can see that decimal constants can have values from -2147483648 to +4294967295. For example, constant number 210 is stored as **unsigned char** (or **unsigned short int**). Similarly, constant number -200 is stored as **signed int**.

Using the suffix u or U forces the constant to be **unsigned**. Using the suffix L or l forces the constant to be **long**. Using both U (or u) and L (or l) forces the constant to be **unsigned long**.

Constants are declared using the keyword **const** and they are stored in the flash program memory of the PIC microcontroller, thus not wasting any valuable RAM space. In the following example, constant **MAX** is declared as 100 and is stored in the flash program memory of the PIC microcontroller:

    const MAX = 100;

Hexadecimal constants start with characters 0x or 0X and may contain numeric data 0 to 9 and hexadecimal characters A to F. In the following example, constant **TOTAL** is given the hexadecimal value FF:

    const TOTAL = 0xFF;

Octal constants have a zero at the beginning of the number and may contain numeric data 0 to 7. In the following example constant **CNT** is given octal value 17:

    const CNT = 017;

Binary constant numbers start with 0b or 0B and may contain only 0 or 1. In the following example, a constant named **Min** is declared having the binary value "11110000":

    const Min = 0b11110000

Floating Point Constants

Floating point constant numbers have integer parts, a dot, fractional part, and an optional e or E followed by a signed integer exponent. In the following example, a constant named **TEMP** is declared having the fractional value 37.50:

    const TEMP = 37.50

or,

    const TEMP = 3.750E1

Character Constants

A character constant is a character enclosed in a single quote. In the following example, a constant named **First_Alpha** is declared having the character value "A":

    const First_Alpha = 'A';

String Constants

String constants are fixed sequences of characters stored in the flash memory of the microcontroller. The string must begin with a double quote character (") and also terminate with a double quote character. The compiler automatically inserts a null character as a terminator. An example string constant is:

    "This is an example string constant"

A string constant can be extended across a line boundary by using a backslash character ("\"):

    "This is first part of the string \
      and this is the continuation of the string"

The above string constant declaration is same as:

    "This is first part of the string and this is the continuation of the string"

Enumarated Constants

Enumaration constants are integer type and they are used to make a program easier to follow. In the following example constant **colours** stores the names of colours. The first element is given the value 0:

    enum colours {black, brown, red, orange, yellow, green, blue, grey, white};

4.2.9 Escape Sequences

Escapes sequences are used to represent non printable ASCII characters. For example, the character combination "\n" represents the newline character. An

ASCII character can also be represented by specifying its hexadecimal code after a backslash. For example, the newline character can also be represented as '\x0A'.

4.2.10 Enumarated Variables

Enumerated variables are used to make a program more readable. In an enumerated variable a list of items is specified and the value of the first item is set to 0, the next item is set to 1, and so on. In the following example type Week is declared as an enumerated list and MON = 0, TUE = 1, WED = 2, and so on):

> **enum** Week {MON, TUE, WED, THU, FRI, SAT, SUN};

It is possible to change the values of the elements in an enumerated list. In the following example black = 2, blue = 3, red = 4 and so on.

> **enum** colours {black = 2, blue, red, white, grey};

Similarly, in the following example, black = 2, blue = 3, red = 8, and grey = 9:

> **enum** colours {black = 2, blue, red = 8, grey};

Variables of type enumeration can be declared by specifying them after the list of items. For example, to declare variable My_Week of enumerated type Week use the following statement:

enum Week {MON, TUE, WED, THU, FRI, SAT, SUN} My_Week;

Now, we can use variable My_Week in our programs:

	My_Week = WED	// assign 2 to My_Week
or		
	My_Week = 2	// same as above

After defining the enumeration type Week, we can declare variables This_Week and Next_Week of type Week as:

> **enum** Week This_Week, Next_Week;

4.2.11 Arrays

Arrays are used to store related items together in the same block of memory and under a specified name. An array is declared by specifying its type, name, and the number of elements it will store. For example,

unsigned int Total[5];

Creates an array of type unsigned int, with name Total, and having 5 elements. The first element of an array is indexed with 0. Thus, in the above example, Total[0] refers to the first element of this array and Total[4] refers to the last element. The array total is stored in memory in five consecutive locations as follows:

| Total[0] |
| Total[1] |
| Total[2] |
| Total[3] |
| Total[4] |

Data can be stored in the array by specifying the array name and index. For example, to store 25 in the second element of the array we have to write:

Total[1] = 25;

Similarly, the contents of an array can be read by specifying the array name and its index. For example, to copy the third array element to a variable called temp we have to write:

Temp = Total[2];

The contents of an array can be initialized during the declaration of the array by assigning a sequence of comma delimited values to the array. An example is given below where array months has 12 elements and months[0] = 31, months[1] = 28, and so on.:

unsigned char months[12] = {31,28,31,30,31,30,31,31,30,31,30,31};

The above array can also be declared without specifying the size of the array:

unsigned char months[ ] = {31,28,31,30,31,30,31,31,30,31,30,31};

Character arrays can be declared similarly. In the following example a character array named Hex_Letters is declared with 6 elements:

unsigned char Hex_Letters[ ] = {'A', 'B', 'C', 'D', 'E', 'F'};

Strings are character arrays with a null terminator. Strings can either be declared by enclosing the string in double quotes, or each character of the array can be specified within single quotes, and then terminated with a null character. In the following example the two string declarations are identical and both occupy 5 locations in memory:

unsigned char Mystring[ ] = "COMP";

and

unsigned char Mystring[ ] = {'C', 'O', 'M', 'P', '\0'};

In C programming language we can also declare arrays with multiple dimensions. One dimensional arrays are usually called vectors, and two dimensional arrays are called matrices. A two dimensional array is declared by specifying the data type of the array, array name, and the size of each dimension. In the following example a two dimensional array named P is created having 3 rows and 4 columns. Altogether the array has 12 elements. The first element of the array is P[0][0], and the last element is P[2][3]. The structure of this array is shown below:

P[0][0]	P[0][1]	P[0][2]	P[0][3]
P[1][0]	P[1][1]	P[1][2]	P[1][3]
P[2][0]	P[2][1]	P[2][2]	P[2][3]

Elements of a multi-dimensional array can be specified during the declaration of the array. In the following example, two dimensional array Q has 2 rows and 2 columns and its diagonal elements are set to one, non-diagonal elements are cleared to zero:

unsigned char Q[2][2] = { {1,0}, {0,1} };

4.2.12 Pointers

Pointers are an important part of the C language and they hold the memory addresses of variables. Pointers are declared same as any other variables, but with the character ("*") in front of the variable name. In general, pointers can be created to point to (or hold the addresses of) character variables, integer variables, long variables, floating point variables, or they can point to functions (mikroC currently does not support pointers to functions).

In the following example, an unsigned character pointer named pnt is declared:

unsigned char *pnt;

When a new pointer is created its content is initially unspecified and it does not hold the address of any variable. We can assign the address of a variable to a pointer using the ("&") character:

pnt = &Count;

now pnt holds the address of variable Count. Variable Count can be set to a value by using the character ("*") in front of its pointer. For example, Count can be set to 10 using its pointer:

*pnt = 10;   // Count = 10

which is same as

Count = 10;   // Count = 10

or, the value of Count can be copied to variable Cnt using its pointer:

Cnt = *pnt;   // Cnt = Count

Array Pointers

In C language the name of an array is also a pointer to the array. Thus, for the array:

unsigned int Total[10];

Name Total is also a pointer to this array and it holds the address of the first element of the array. Thus, the following two statements are equal:

Total[2] = 0;
and
*(Total + 2) = 0;

Also, the following statement is true:

&Total[j] = Total + j

In C language we can perform pointer arithmetic which may involve:

- Comparing two pointers
- Adding or subtracting pointer and an integer value
- Subtracting two pointers
- Assigning one pointer to another one
- Comparing a pointer to null

For example, let us assume that pointer P is set to hold the address of array element Z[2]

P = &Z[2];

We can now clear elements 2 and 3 of array Z as in the following two examples. The two examples are identical except that in the first example pointer P holds the address of Z[3] at the end of the statements, and it holds the address of Z[2] at the end of the second set of statements :

```
*P = 0;              // Z[2] = 0
P = P + 1;           // P now points to element 3 of Z
*P = 0;              // Z[3] = 0
```
or
```
*P = 0;              // Z[2] = 0
*(P + 1) = 0;        // Z[3] = 0
```

A pointer can be assigned to another pointer. An example is given below where variables Cnt and Tot are both set to 10 using two different pointers:

```
unsigned int *i, *j;     // declare 2 pointers
unsigned int Cnt, Tot;   // declare two variables
i = &Cnt;                // i points to Cnt
*i = 10;                 // Cnt = 10
j = i;                   // copy pointer i to pointer j
Tot = *j;                // Tot = 10
```

4.2.13 Structures

Structures can be used to collect related items as single objects. Unlike arrays, the members of structures can be a mixture of any data type. For example, a structure can be created to store the personal details (name, surname, age, date of birth etc.) of a student.

A structure is created by using the keyword **struct**, followed by a structure name, and a list of member declarations. Optionally, variables of the same type as the structure can be declared at the end of the structure.

The following example declares a structure named Person:

```
struct Person
{
    unsigned char name[20];
    unsigned char surname[20];
    unsigned char nationality[20];
    unsigned char age;
}
```

Declaring a structure does not occupy any space in memory, but the compiler creates a template describing the names and types of the data objects or member elements that will eventually be stored within such a structure variable. It is only when variables of the same type as the structure are created then these variables occupy space in memory. We can declares variables of the same type as the structure by giving the name of the structure and the name of the variable. For example, two variables **Me** and **You** of type Person can be created by the statement:

struct Person Me, You;

Variables of type Person can also be created during the declaration of the structure as shown below:

```
struct Person
{
    unsigned char name[20];
    unsigned char surname[20];
    unsigned char nationality[20];
    unsigned char age;
} Me, You;
```

We can assign values to members of a structure by specifying the name of the structure, followed by a dot ("."), and the name of the member. In the following example, the **age** of structure variable **Me** is set to 25, and variable **M** is assigned to the value of **age** in structure variable **You**:

Me.age = 25;
M = You.age;

Structure members can be initialized during the declaration of the structure. In the following example, the radius and height of structure Cylinder are initialized to 1.2 and 2.5 respectively:

```
struct Cylinder
{
        float radius;
        float height;
} MyCylinder = {1.2, 2.5};
```

Values can also be set to members of a structure using pointers by defining the variable types as pointers. For example, if **TheCylinder** is defined as a pointer to structure Cylinder then we can write:

```
struct Cylinder
{
        float radius;
        float height;
} *TheCylinder;

TheCylinder -> radius = 1.2;
TheCylinder -> height = 2.5;
```

The size of a structure is the number of bytes contained within the structure. We can use the **sizeof** operator to get the size of a structure. Considering the above example,

sizeof(MyCylinder)

returns 8 since each float variable occupies 4 bytes in memory.

Bit fields can be defined using structures. With bit fields we can assign identifiers to bits of a variable. For example, to identify bits 0, 1, 2 and 3 of a variable as **LowNibble** and to identify the remaining 4 bits as **HighNibble** we can write:

```
struct
{
        LowNibble  : 4;
        HighNibble : 4;
```

} MyVariable;

We can then access the nibbles of variable MyVariable as:

MyVariable.LowNibble = 12;
MyVariable.HighNibble = 8;

In C language we can use the **typedef** statements to create new types of variables. For example, a new structure data type named **Reg** can be created as follows:

typedef struct
{
    unsigned char name[20];
    unsigned char surname[20];
    unsigned age;
} Reg;

Variables of type **Reg** can then be created in exactly the same way as creating any other types of variables. In the following example, variables **MyReg**, **Reg1** and **Reg2** are created from data type **Reg**:

Reg MyReg, Reg1, Reg2;

The contents of one structure can be copied to another structure, provided that both structures have been derived from the same template. In the following example two structure variables P1 and P2 of same type have been created and P2 is copied to P1:

struct Person
{
    unsigned char name[20];
    unsigned char surname[20];
    unsigned int age;
    unsigned int height;
    unsigned weight;
}

struct Person P1, P2;
....
....
P2 = P1;

4.2.14 Operators in C

Operators are applied to variables and other objects in expressions and they cause some conditions or some computations to occur.

mikroC language supports the following operators:

- Arithmetic operators
- Logical operators
- Bitwise operators
- Conditional operators
- Assignment operators
- Relational operators
- Pre-processor operators

Arithmetic Operators

Arithmetic operators are used in arithmetic computations. Arithmetic operators associate from left to right and they return numerical results. A list of the mikroC arithmetic operators is given in Table 4.3.

Table 4.3 mikroC arithmetic operators

Operator	Operation
+	Addition
-	Subtraction
*	Multiplication
/	Division
%	Remainder (integer division)
++	Auto increment
--	Auto decrement

Example use of arithmetic operators is given below:

```
/* Adding two integers */
5 + 12                          // equals 17

/* Subtracting two integers */
120 – 5                         // equals 115
10 – 15                         // equals -5

/* Dividing two integers */
```

5 / 3	// equals 1
12 / 3	// equals 4

```
/* Multiplying two integers */
3 * 12                                  // equals 36

/* Adding two floating point numbers */
3.1 + 2.4                               // equals 5.5

/* Multiplying two floating point numbers */
2.5 * 5.0                               / equals 12.5

/* Dividing two floating point numbers */
25.0 / 4.0                              // equals 6.25

/* Remainder (not for float) */
7 % 3                                   // equals 1

/* Post-increment operator */
j = 4;
k = j++;                                // k = 4, j = 5

/* Pre-increment operator */
j = 4;
k = ++j;                                // k = 5, j = 5

/* Post-decrement operator */
j = 12;
k = j--;                                // k = 12, j = 11

/* Pre-decrement operator */
j = 12;
k = --j;                                // k = 11, j = 11
```

Relational Operators

Relational operators are used in comparisons. If the expression evaluates to TRUE, a 1 is returned, otherwise a 0 is returned.

All relational operators associate from left to right and a list of mikroC relational operators is given in Table 4.4.

Table 4.4 mikroC relational operators

Operator	Operation
==	Equal to
!=	Not equal to
>	Greater than
<	Less than
>=	Greater than or equal to
<=	Less than or equal to

Logical Operators

Logical operators are used in logical and arithmetic comparisons and they return TRUE (i.e. logical 1) if the expression evaluates to nonzero, or FALSE (i.e. logical 0) if the expression evaluates to zero. If more than one logical operator is used in a statement and if the first condition evaluates to false, the second expression is not evaluated.

A list of the mikroC logical operators is given in Table 4.5.

Table 4.5 mikroC logical operators

Operator	Operation
&&	AND
||	OR
!	NOT

Pre-Processor Operators

The pre-processor allows a programmer to:

- Compile a program conditionally such that parts of the code is not compiled
- Replace symbols with other symbols or values
- Insert text files into a program

The pre-processor operator is the ("#") character and any line of code with a leading ("#") is assumed to be a pre-processor command. Semicolon character (";") is not needed to terminate a pre-processor command.

mikroC compiler supports the following pre-processor commands:

#define   #undef
#if   #elif   #endif
#ifdef   #ifndef
#error
#line

#define, #undef, #ifdef, #ifndef

#define pre-processor command provides Macro expansion where every occurrence of an identifier in the program is replaced with the value of the identifier. For example, to replace every occurrence of MAX with value 100 we can write:

#define MAX 100

An identifier which has already been defined can not be defined again unless both definitions have the same values. One way to get round this problem is to remove the Macro definition:

#undef MAX

or, the existence of a Macro definition can be checked. In the following example, if **MAX** has not already been defined then it is given value 100, otherwise the #define line is skipped:

#ifndef MAX
    #define MAX 100
#endif

Note that the **#define** pre-processor command does not occupy any space in memory.

We can pass parameters to a Macro definition by specifying the parameters in a parenthesis after the Macro name. For example, consider the Macro definition

#define ADD(a, b)(a + b)

when this Macro is used in a program, (a,b) will be replaced with (a + b) as shown below:

p = ADD(x, y)   will be transformed into p = (x + y)

Similarly, we can define a Macro to calculate the square of two numbers:

#define SQUARE(a)   (a * a)

when we now use this Macro in a program:

p = SQUARE(x)   will be transformed into p = (x * x)

#include

The pre-processor directive #include is used to include a source file in our program. Usually header files with extension ".h" are used with #include. There are two formats of using the **#include**:

#include <file>

and

#include "file"

In first option the file is searched in the mikroC installation directory first and then in user search paths. In second option the specified file is searched in the mikroC project folder, then in the mikroC installation folder, and then in user search paths. It is also possible to specify a complete directory path as:

#include "C:\temp\last.h"

The file is then searched only in the specified directory path.

4.2.15 Modifying the flow of control

Statements are normally executed sequentially from the beginning to the end of a program. We can use control statements to modify the normal sequential flow of control in a C program. The following control statements are available in mikroC programs:

- Selection statements
- Unconditional modification of flow
- Iteration statements

Selection Statements

There are two selection statements: *If* and *switch*.

If Statement

The general format of the **if** statement is:

    if(expression)
        Statement1;
    else
        Statement2;

or,

    if(expression)Statement1; **else** Statement2;

If the **expression** evaluates to TRUE, **Statement1** is executed, otherwise **Statement2** is executed. The **else** keyword is optional and may be omitted if not required. In the following example, if the value of **x** is greater than **MAX** then variable **P** is incremented by 1, otherwise it is decremented by 1:

    if(x > MAX)
        P++;
    else
        P--;

We can have more than one statement by enclosing the statements within curly brackets. For example,

    if(x > MAX)
    {
        P++;
        Cnt = P;
        Sum = Sum + Cnt;
    }
    else
        P--;

In the above example if **x** is greater than **MAX** then the three statements within the curly brackets are executed, otherwise the statement **P--** is executed.

Another example using the **if** statement is given below:

    if(x > 0 && x < 10)
    {
        Total += Sum;
        Sum++;
    }

```
else
{
        Total = 0;
        Sum = 0;
}
```

switch Statement

The **switch** statement is used when there are a number of conditions and different operations are performed when a condition is true. The syntax of the **switch** statement is:

```
switch (condition)
{
        case condition1:
                Statements;
                break;
        case condition2:
                Statements;
                break;
        .....................
        .....................
        case conditionn:
                Statements;
                break;
        default:
                Statements;
}
```

The **switch** statement functions as follows: First the **condition** is evaluated. The **condition** is then compared to **condition1** and if a match is found statements in that case block are evaluated and control jumps outside the **switch** statement when the **break** keyword is encountered. If a match is not found, **condition** is compared to **condition2** and if a match is found statements in that case block are evaluated and control jumps outside the switch statements and so on. The **default** is optional and statements following **default** are evaluated if the **condition** does not match to any of the conditions specified after the **case** keywords.

In the following example, the value of variable Cnt is evaluated. If Cnt = 1, A is set to 1. If cnt = 10, B is set to 1, and if Cnt = 100, C is set to 1. If Cnt is not equal to 1, 10, or 100 then D is set to 1:

```
switch (Cnt)
{
    case 1:
        A = 1;
        break;
    case 10:
        B = 1;
        break;
    case 100:
        C = 1;
        break;
    default:
        D = 1;
}
```

Because white spaces are ignored in C language we could also write the above code as:

```
switch (Cnt)
{
    case 1:     A = 1;break;
    case 10:    B = 1;break;
    case 100:   C = 1;break;
    default:    D = 1;
}
```

ITERATION STATEMENTS

Iteration statements enable us to perform loops in our programs where part of a code is repeated required number of times. In mikroC there are 4 ways that iteration can be performed and we will look at each one with examples:

- Using **for** statement
- Using **while** statement
- Using **do** statement
- Using **goto** statement

for Statement

The syntax of the **for** statement is:

```
for(initial expression; condition expression; increment expression)
{
        Statements;
}
```

The **initial expression** sets the starting variable of the loop and this variable is compared against the **condition expression** before an entry to the loop. Statements inside the loop are executed repeatedly, and after each iteration the value of **increment expression** is incremented. The iteration continues until the **condition expression** becomes false. An endless loop is formed if the **condition expression** is always true.

The following example shows how a loop can be set up to execute 10 times. In this example variable **i** starts from 0 and increments by 1 at the end of each iteration. The loop terminates when i =10 in which case the condition i < 10 becomes false. On exit from the loop the value of **i** is 10:

```
for(i = 0; i < 10; i ++)
{
        statements;
}
```

The above loop could also be formed by starting the **initial expression** with a nonzero value. Here, **i** starts with 1 and the loop terminates when I = 11. Thus, on exit from the loop the value of **i** is 11:

```
for(i = 1; i <= 10; i++)
{
        Statements;
}
```

The parameters of a **for** loop are all optional and can be omitted. If the **condition expression** is left out, it is assumed to be true. In the following example an endless loop is formed where the **condition expression** is always true and the value of **i** is starts with 0 and is incremented after each iteration:

```
/* Endless loop with incrementing i */
for(i=0; ; i++)
{
        Statements;
}
```

Another example of an endless loop is given below where all the parameters are omitted:

```
/* Example of endless loop */
for(; ;)
{
    Statements;
}
```

In the following endless loop **i** starts with 1 and is not incremented inside the loop:

```
/* Endless loop with i = 1 */
for(i=1; ;)
{
    Statements;
}
```

If there is only one statement inside the **for** loop we can omit the curly brackets as shown in the following example:

for(k = 0; k < 10; k++)Total = Total + Sum;

Nested for loops can be used in programs. In a nested for loop the inner loop is executed for each iteration of the outer loop. An example is given below where the inner loop is executed 5 times and the outer loop is executed 10 times. The total iteration count is 50:

```
/* Example of nested for loops */
for(i = 0; i < 10; i++)
{
    for(j = 0; j < 5; j++)
    {
        Statements;
    }
}
```

In the following example the sum of all the elements of a 3x4 matrix M is calculated and stored in variable called Sum:

```
/* Add all elements of a 3x4 matrix */
Sum = 0;
for(i = 0; i < 3; i++)
```

```
        {
                for(j = 0; j < 4; j++)
                {
                        Sum = Sum + M[i][j];
                }
        }
```

Since there is only one statement to be executed, the above example could also be written as:

```
/* Add all elements of a 3x4 matrix */
Sum = 0;
for(i = 0; i < 3; i++)
{
        for(j = 0; j < 4; j++) Sum = Sum + M[i][j];
}
```

while Statement

This is another statement which can be used to create iteration in programs. The syntax of the **while** statement is:

```
while (condition)
{
        Statements;
}
```

Here, the statements are executed repeatedly until the **condition** becomes false, or, the statements are executed repeatedly as long as the **condition** is true. If the **condition** is false on entry to the loop then the loop will not be executed and the program will continue from the end of the **while** loop. It is important that the **condition** is changed inside the loop, otherwise an endless loop will be formed.

The following code shows how to set up a loop to execute 10 times using the **while** statement:

```
/* A loop that executes 10 times */
k = 0;
while (k < 10)
{
        Statements;
        k++;
}
```

At the beginning of the code variable **k** is 0. Since **k** is less than 10 the **while** loop starts. Inside the loop the value of **k** is incremented by 1 after each iteration. The loop repeats as long as k < 10 and is terminated when k = 10. At the end of the loop the value of **k** is 10.

Notice that an endless loop will be formed if **k** is not incremented inside the loop:

```
/* An endless loop */
k = 0;
while (k < 10)
{
        Statements;
}
```

An endless loop can also be formed by setting the **condition** to be always true:

```
/* An endless loop */
while (k = k)
{
        Statements;
}
```

Here is an example of calculating the sum of numbers from 1 to 10 and storing the result in variable called **sum**:

```
/* Calculate the sum of numbers from 1 to 10 */
unsigned int k, sum;
k = 1;
sum = 0;
while(k <= 10)
{
        sum = sum + k;
        k++;
}
```

It is possible to have a **while** statement with no body. Such a statement is useful for example if we are waiting for an input port to change its value. An example is given below where the program will wait as long as bit 0 of PORT B (PORTB.0) is at logic 0. The program will continue when the port pin changes to logic 1:

```
while(PORTB.0 == 0);        // Wait until PORTB.0 to becomes 1
```

or,

    while(PORTB.0);

It is possible to have nested **while** statements.

do **Statement**

The **do** statement is similar to the **while** statement but here the loop executes until the **condition** becomes false, or, the loop executes as long as the **condition** is true. The **condition** is tested at the end of the loop. The syntax of the **do** statement is:

```
do
{
    Statements;
} while (condition);
```

The first iteration is always performed whether the **condition** is true or false, and this is the main difference between the **while** statement and the **do** statement.

The following code shows how to setup a loop to execute 10 times using the **do** statement:

```
/* Execute 10 times */
k = 0;
do
{
    Statements;
    k++;
} while (k < 10);
```

The loop starts with k = 0 and the value of **k** is incremented inside the loop after each iteration. **k** is tested at the end of the loop and if **k** is not less than 10 the loop terminates. In this example because k = 0 at the beginning of the loop, the value of **k** is 10 at the end of the loop.

An endless loop will be formed if the condition is not modified inside the loop as shown in the following example. Here **k** is always less than 10:

```
/* An endless loop */
k = 0;
do
{
    Statements;
} while (k < 10);
```

An endless loop can also be created if the condition is set to be true all the time:

```
/* An endless loop */
do
{
    Statements;
} while (k = k);
```

It is possible to have nested **do** statements.

goto Statement

The **goto** statement can be used to alter the normal flow of control in a program. This statement causes the program to jump to a specified label. A label can be any alphanumeric character set starting with a letter and terminating with the colon (":") character.

Although not recommended, the **goto** statement can be used together with the **if** statement to create iterations in a program. The following example shows how to setup a loop to execute 10 times using the **goto** and **if** statements:

```
      /* Execute 10 times */
      k = 0;
Loop:
      Statements;
      k++;
      if(k < 10)goto Loop;
```

The loop starts with label **Loop** and variable k = 0 at the beginning of the loop. Inside the loop the statements are executed and **k** is incremented by 1. The value of **k** is then compared with 10 and the program jumps back to label **Loop** if k < 10. Thus, the loop is executed 10 times until the condition at the end becomes false. At the end of the loop the value of **k** is 10.

4.3 PIC Microcontroller Input-Output Port Programming

Depending on the type of microcontroller used, PIC microcontroller input-output ports are named as PORTA, PORTB, PORTC and so on. Port pins can be in analog or digital mode. In analog mode ports are input only and a built-in analog to digital converter and multiplexer circuits are used. In digital mode a port pin can either be configured as input or output. The TRIS registers control the port directions and there are TRIS registers for each port, named as TRISA, TRISB, TRISC and so on. Clearing a TRIS register bit to 0 sets the corresponding port bit to output mode. Similarly, setting a TRIS register bit to 1 sets the corresponding port bit to input mode.

Ports can be accessed as either a single 8-bit register, or individual bits of a port can be accessed. In the following example PORT B is configured as an output port and all its bits are set to a 1:

    TRISB = 0;   // Set PORT B as output
    PORTB = 0xFF;   // Set PORTB bits to 1

Similarly, the following example shows how 4 upper bits of PORT C can be set as input and how upper 4 bits of PORT C can be set as output:

    TRISC = 0xF0;

Bits of an input-output port can be accessed by specifying the required bit number. In the following example, variable P2 is loaded with bit 2 of PORT B:

    P2 = PORTB.2;

All the bits of a port can be complemented by the statement:

    PORTB = ~PORTB;

4.4 Programming Examples

In this section some simple programming examples are given to make the reader familiar with programming in C.

Example 4.1

Write a program to set all 8 port pins of PORT B to logic 1.

Solution 4.1

The required program is given below. PORT B is configured as an output port and then all port pins are set to logic 1 by sending hexadecimal number 0xFF:

```
void main()
{
    TRISB = 0;           // Configure PORT B as output
    PORTB = 0xFF;        // Set all port pins to logic a
}
```

Example 4.2

Write a program to set the odd numbered (bits 1, 3, 5, and 7) PORT B pins to logic 1.

Solution 4.2

Odd numbered port pins can be set to logic 1 by sending the bit pattern "10101010" to the port. This bit pattern is the hexadecimal number 0xAA and the required program is:

```
void main()
{
    TRISB = 0;           // Configure PORT B as output
    PORTB = 0xAA;        // Turn on odd numbered port pins
}
```

Example 4.3

It is required to write a program to continuously count up in binary and send this data to PORT B. Thus, PORT B is required to have the binary data:

```
00000000
00000001
00000010
00000011
............
............
11111110
11111111
00000000
............
```

68

Solution 4.3

A **for** loop can be used to create an endless loop and inside this loop the value of a variable can be incremented and then sent to PORT B. The required program is:

```
void main()
{
    unsigned char Cnt = 0;

    for(;;)                    // Endless loop
    {
        PORTB = Cnt;           // Send Cnt to PORT B
        Cnt++;                 // Increment Cnt
    }
}
```

Example 4.4

Write a program to set all bits of PORT B to logic 1 and then to logic 0. Repeat this process 10 times.

Solution 4.4

The **for** statement can be used to create a loop and repeat the required operation 10 times:

```
void main()
{
    unsigned char j;

    for(j = 0; j < 10; j++)    // Repeat 10 times
    {
        PORTB = 0xFF;          // Set PORT B pins to 1
        PORTB = 0;             // Clear PORT B pins
    }
}
```

Example 4.5

The radius and height of a cylinder are 2.5cm and 10cm respectively. Write a program to calculate the volume of this cylinder.

Solution 4.5

The required program is:

```
void main()
{
    float Radius = 2.5, Height = 10;
    float Volume;

    Volume = PI *Radius*Radius*Height;
}
```

Example 4.6

Write a program to find the largest element of an integer array having 10 elements.

Solution 4.6

The program is given below. At the beginning variable **m** is set to the first element of the array. A loop is then formed and the largest element of the array is found:

```
void main()
{
    unsigned char j;
    int m, A[10];

    m = A[0];                    // First element of array
    for(j = 1; j < 10; j++)
    {
        if(A[j] > m)m = A[j];
    }
}
```

Example 4.7

Write a program using the **while** statement to clear all 10 elements of an integer array M.

Solution 4.7

As shown in the program listing below, **NUM** is defined to be 10 and variable **j** is used as the loop counter:

```
#define NUM 10
void main()
{
        int M[NUM];
        unsigned char j = 0;

        while (j < NUM)
        {
                M[j] = 0;
                j++;
        }
}
```

Example 4.8

Write a program to convert the temperature from °C to °F starting from 0°C, in steps of 1°C up to and including 100°C, and store the results in an array called F.

Solution 4.8

Given the temperature in °C, the equivalent in °F is calculated using the formula:

$$F = (C - 32.0) / 1.8$$

The required program listing is given below. A for loop is used to calculate the temperature in °F and store in array F:

```
void main()
{
        float F[100];
        unsigned char C;

        for(C =0; C <= 100; C++)
        {
                F[C] = (C - 32.0) / 1.8;
        }
}
```

CHAPTER 5

THE *EasyPIC 5* MICROCONTROLLER DEVELOPMENT BOARD

5.1 Overview

The EasyPIC 5 microcontroller development board (see Figure 5.1) is developed and manufactured by mikroElektronika (www.mikroe.com). This is a highly popular microcontroller development board that can be used to build small to medium size and even complex projects. Power to the board can either be supplied from the USB port by connecting to a PC, or an external power supply can be used. The board incorporates an on-board chip programmer and an in-circuit debugger. Further details of the EasyPIC 5 development board are given in this chapter together with simple projects.

Figure 5.1 EasyPIC 5 microcontroller development board

5.2 Specification of the EasyPIC 5 Development Board

The specifications of the EasyPIC 5 microcontroller development board are given below:

- Support for 8,14,18,20,28 and 40 pin microcontrollers
- On board USB 2.0 programmer
- On board in-circuit debugging
- 36 LEDs
- 36 buttons
- RS232 communications port
- 7-segment multiplexed display
- DS1820 temperature sensor
- 2 potentiometers for A/D conversion testing
- DIP switches for port pull-up and pull-down
- All port pins connected to IDC connectors
- Text based LCD support
- Graphics based LCD support
- On board touch screen controller
- Crystal can be removed and replaced
- RESET circuit
- PS2 connector
- Various jumpers to configure the board

The EasyPIC 5 development board is ready to use after installing the PICflash programmer software (see *PICflash programmer manual* or the CDROM delivered with the board) and the USB drivers (see *Installing USB drivers manual* or the CDROM delivered with the board) on the PC.

The EasyPIC 5 development board can easily be used with the mikroC, mikroBASIC and mikroPASCAL language compilers. In this book the mikroC language is used in all projects. Next section gives some simple microcontroller projects based on using the mikroC language together with the EasyPIC 5 microcontroller development board.

5.3 Simple Projects With the EasyPIC 5 Development Board

In this section the design of several PIC16F887 based simple projects are given. All the projects are developed using the mikroC language compiler and the

EasyPIC 5 microcontroller development board. The following information is given for each project:

- Project title
- Brief description of the project
- Block diagram of the project
- Circuit diagram and EasyPIC 5 jumper settings for the project
- Program Description Language (PDL) of the project
- Program listing and description
- Suggestions for possible improvement of the project

Before describing the projects it is worthwhile to have a look at how the PDL of a project can be developed as it will be used to describe the operation of each project.

A PDL consists of simple English-like statements used to describe the operation of a program code. PDL is an alternative to the classical flow-chart and offers the advantages that it is simpler than a flow-chart as there is no need to draw boxes. The PDL is free format and most of the commonly used PDL statements are given in the following sub-sections.

5.3.1 BEGIN – END

This pair of statements are used to indicate the start and end of a code. Every PDL code starts with a "BEGIN" and terminates with an "END" as shown below (words "BEGIN" and "END" are usually written in bold and statements inside are indented for clarity):

BEGIN
....
....
....
END

5.3.2 Statements

The statements describe the actual operations to be performed by the code and they are written in free format using plain English sentences. An example is given below where it is required to turn the LED ON for two seconds and then to turn in OFF:

BEGIN
    Turn ON LED
    Wait 2 seconds
    Turn OFF LED
END

5.3.3 IF – ELSE – ENDIF

These conditional statements are used for the control of the flow of execution. The code inside an IF-ELSE-ENDIF statement is usually indented for clarity. An example is give below:

BEGIN
    Turn ON LED 1
    Wait for 5 seconds
    IF Switch is pressed
        Turn OFF LED 1
        Wait for 2 seconds
    ELSE
        Turn ON LED 2
    ENDIF
END

5.3.4 REPEAT – UNTIL

This flow of control construct is used to repeat part of a code until a certain condition is satisfied. Statements "REEPAT" and "UNTIL" are usually written in bold and the code inside in indented for clarity. An example is given below where in this example initially all LEDs are turned OFF. Then LED 1 is then flashed with one second intervals. The flashing is repeated until a button is pressed:

BEGIN
    Turn OFF all LEDs
    REPEAT
        Turn ON LED 1
        Wait 1 second
        Turn OFF LED 1
        Wait 1 second
    UNTIL button is pressed

END

5.3.5 DO FOREVER – ENDDO

The statements between the "DO FOREVER and "ENDDO" are repeated indefinitely in a loop. The statements inside the construct are indented for clarity. An example is given below where the LED is flashed continuously with one second interval:

    BEGIN
        Turn OFF all LEDs
        DO FOREVER
            Turn ON LEDs
            Wait 1 second
            Turn OFF LEDs
            Wait 1 second
        ENDDO
    END

5.3.6 DO n TIMES - ENDDO

The statements between the "FOR" and "END" are repeated in a loop as many times as specified by the heading of the loop. An example is given below where all the LEDare initially turned OFF and then they are flashed 5 times:

    BEGIN
        Turn OFF all LEDs
        DO 5 TIMES
            Turn ON all LEDs
            Wait 1 second
            Turn OFF all LEDs
            Wait 1 second
        ENDDO
    END

5.3.7 WHILE – END

The statements between "WHILE" and "END" are repeated while a given condition is true. An example is given below where all the LEDs are turned OFF initially and then they are flashed 5 times with 1 second interval:

BEGIN
    Turn OFF all LEDs
    Set J = 0
    WHILE J < 5
        Turn ON all LEDs
        Wait 1 second
        Turn OFF all LEDs
        Wait 1 second
        Increment J
    END
END

5.4 PROJECT 1 – FLASHING LEDs

5.4.1 Description

In this project 8 LEDs are connected to PORT C of a PIC16F887 microcontroller. The LEDs are turned ON and OFF with one second intervals.

5.4.2 Block Diagram

The block diagram of the project is shown in Figure 5.2.

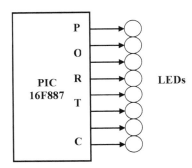

Figure 5.2 Block diagram of the project

5.4.3 Circuit Diagram

The circuit diagram of the project is shown in Figure 5.3. The LEDs are connected to PORT C pins via current limiting resistors. The value of the resistors can be found as follows:

$$R = (V_D - V_f) / I_f$$

Where, V_D is the supply voltage (5V)
   V_f is the LED forward voltage (2V)
   I_f is the LED forward current (typically 10mA)

Thus, $R = (5 - 2) / 10 = 300$ ohm. We can use a 390 ohm (or a 470 ohm) resistor.

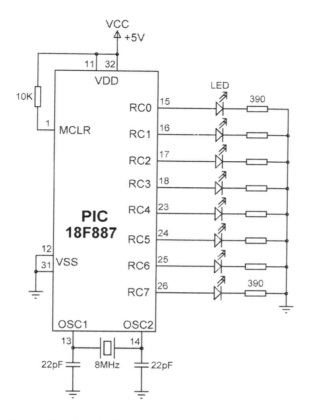

Figure 5.3 Circuit diagram of the project

5.4.4 PDL of the Project

The PDL of the project is shown in Figure 5.4. The LEDs are flashed continuously with an interval of one second.

       BEGIN
          Configure PORT C pins as outputs
          DO FOREVER
              Turn ON PORT C LEDs
              Wait 1 second
              Turn OFF PORT C LEDs
              Wait 1 second
          ENDDO

END

Figure 5.4 PDL of the project

5.4.5 Program Listing

The program listing of the project is shown in Figure 5.5. At the beginning of the project PORT C pins are configured as outputs using the statement TRISC = 0. Then an endless **for** loop is formed where the PORT C pins are turned ON and OFF with one second intervals. The mikroC built-in function **Delay_Ms(1000)** is used to create the required one second delay.

```
/*************************************************************
                    LED FLASHING PROGRAM
                    ========================

In this project 8 LEDs are connected to PORT C of a PIC16F887
microcontroller. The program flashes all the LEDs with one second interval.

The project is built on an EasyPIC 5 development kit, operating with a
8MHz crystal.


    Programmer      : Dogan Ibrahim
    File            : PROJECT1.C
    Microcontroller : PIC16F887
*************************************************************/

void main()
{
    TRISC = 0;                  // PORT C pins are outputs
    for(;;)                     // Endless loop
    {
        PORTC = 0;              // LEDs are OFF
        Delay_Ms(1000);         // Wait 1 second
        PORTC = 0xFF;           // LEDs are ON
        Delay_Ms(1000);         // Wait 1 second
    }
}
```

Figure 5.5 Program listing of the project

The steps to create the program, load it to the EasyPIC 5 development kit and the jumper configurations are given in the following steps:

- Start the mikroC compiler

- Select **Project -> New Project**
- Select a project name (**Project1** in this example)
- Select the device as PIC16F887, clock as 8 MHz, and set the following device flags (high frequency clock, watchdog OFF, and low-voltage programming OFF) as shown in Figure 5.6:

> HS_OSC
> WDT_OFF
> LVP_OFF

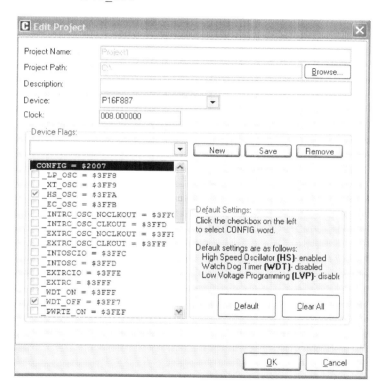

Figure 5.6 Select the device and device flags

- Write the program as in Figure 5.5 (see Figure 5.7)

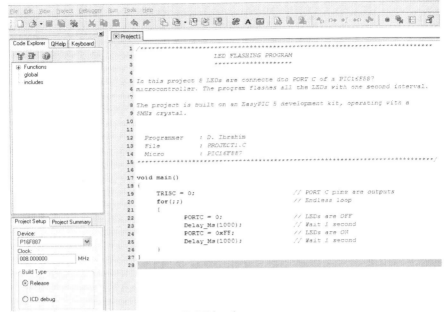

Figure 5.7 Write the program

- Compile the program by clicking the Build Project as in Figure 5.8

Figure 5.8 Compile the program

- Connect the EasyPIC 5 development kit to the PC via the supplied USB cable. The orange USB LINK LED on the board should turn ON. and select the following jumpers (see Figure 5.9):

        SUPPLY SELECT: USB
        SW6   :   PORT C LEDs ON

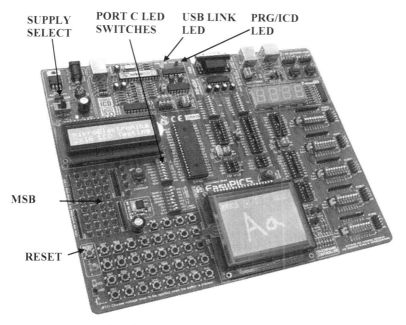

Figure 5.9 Jumper and LED positions on EasyPIC 5

- Load the microcontroller program memory by clicking **Tools -> me Programmer**. The PRG/ICD LED should flash on the board while the device is programmed.
- The program will be loaded into the microcontroller and PORT C LEDs will flash as required (Press the RESET button if required) with one second intervals.

Modification to the Program

The program given in Figure 5.5 can be made more friendly and easier to understand if **"define"** statements are used to define the states of the LEDs. The modified program (PROJECT1-1.C) is shown in Figure 5.10.

```
/*********************************************************
                    LED FLASHING PROGRAM
                    ************************
```

In this project 8 LEDs are connected to PORT C of a PIC16F887 microcontroller. The program flashes all the LEDs with one second interval.

The project is built on an EasyPIC 5 development kit, operating with a 8MHz crystal.

In this program the "define" statement is used to make the program easier to understand

```
    Programmer      : Dogan Ibrahim
    File            : PROJECT1-1.C
    Microcontroller : PIC16F887
*********************************************************/
#define OFF 0
#define ON  0xFF
#define LEDS PORTC

void main()
{
   TRISC = 0;                   // PORT C pins are outputs
   for(;;)                      // Endless loop
   {
      LEDS = OFF;               // LEDs are OFF
      Delay_Ms(1000);           // Wait 1 second
      LEDS = ON;                // LEDs are ON
      Delay_Ms(1000);           // Wait 1 second
   }
}
```

Figure 5.10 Modified program

The one second delay routine can also be declared using a "define" statement. This is shown in Figure 5.11 (PROJECT1-2.C).

```c
/*******************************************************************
             LED FLASHING PROGRAM
             ************************

In this project 8 LEDs are connected to PORT C of a PIC16F887
microcontroller. The program flashes all the LEDs with one second interval.

The project is built on an EasyPIC 5 development kit, operating with a
8MHz crystal.

In this modified version of the program, the delay routine is defined as a Macro


  Programmer       : Dogan Ibrahim
  File             : PROJECT1-2.C
  Microcontroller  : PIC16F887
*******************************************************************/
#define OFF 0
#define ON  0xFF
#define LEDS PORTC
#define Wait_One_Second Delay_Ms(1000)

void main()
{
   TRISC = 0;                     // PORT C pins are outputs
   for(;;)                        // Endless loop
   {
       LEDS = OFF;                // LEDs are OFF
       Wait_One_Second;           // Wait 1 second
       LEDS = ON;                 // LEDs are ON
       Wait_One_Second;           // Wait 1 second
   }
}
```

Figure 5.11 Using **define** statement for the delay routine

Longer Delay

In some applications it may be required to have longer delays. The program code in Figure 5.12 (PROJECT1-3.C) shows how a function can be used to introduce variable delays. Here, 5 second delay is introduced between each output by calling to user function **Variable_Delay**.

5.4.6 Suggestions for Future Work

The project can be modified such that LEDs on other ports can be turned ON and OFF. For example, you may like to try flashing LEDs on both PORT B and PORT C.

```
/*************************************************************
                    LED FLASHING PROGRAM
                    ************************

In this project 8 LEDs are connected to PORT C of a PIC16F887
microcontroller. The program flashes all the LEDs with one second interval.

The project is built on an EasyPIC 5 development kit, operating with a
8MHz crystal.

In this version of the program a function is used to introduce delay to the program


        Programmer        : Dogan Ibrahim
        File              : PROJECT1-3.C
        Microcontroller   : PIC16F887
*************************************************************/
#define OFF 0
#define ON  0xFF
#define LEDS PORTC

//
// The following function introduces variable delay in one second units
//
Variable_Delay(unsigned int d)
{
   unsigned int i;

   for(i = 0; i < d; i++)Delay_Ms(1000);
}

void main()
{
   TRISC = 0;                       // PORT C pins are outputs
   for(;;)                          // Endless loop
   {
        LEDS = OFF;                 // LEDs are OFF
        Variable_Delay(5);          // Wait 5 seconds
        LEDS = ON;                  // LEDs are ON
        Variable_Delay(5);          // Wait 5 seconds
   }
}
```

Figure 5.12 Using a function to introduce delay

5.5 PROJECT 2 – MOVING LEDs

5.5.1 Description

In this project 8 LEDs are connected to PORT C of a PIC16F887 microcontroller as in Project 1. As shown below, the LEDs are turned ON and OFF alternatively and with one second intervals to give the impression that they are moving:

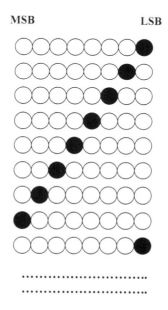

5.5.2 Block Diagram

The block diagram of the project is as shown in Figure 5.2.

5.5.3 Circuit Diagram

The circuit diagram of the project is as shown in Figure 5.3.

5.5.4 PDL of the Project

The PDL of the project is shown in Figure 5.13. The LEDs are flashed continuously with an interval of one second.

```
           BEGIN
                Configure PORT C pins as outputs
                Set Count = 1
                DO FOREVER
                     Send Count to PORT C
                     Wait 1 second
                     Shift Count left by 1 digit
                     IF Count = 0
                          Count = 1
                     ENDIF
                ENDDO
           END
```

Figure 5.13 PDL of the project

5.5.5 Program Listing

The program listing of the project (PROJECT2.C) is shown in Figure 5.14. At the beginning of the project PORT C pins are configured as outputs using the statement TRISC = 0. Then, character variable **Count** is set to 1 and an endless **for** loop is formed. Inside this loop the value of **Count** is left shifted by 1 digit and then sent to PORT C with 1 second intervals. When the MSB LED is turned ON the value of **Count** is 128. Shifting **Count** left by one digit will cause overflow and will clear **Count** to 0. This is detected by the IF condition and variable **Count** is set back to 1. The process is repeated forever.

```
/***********************************************************
                    MOVING LEDS
                  **************
```

In this project 8 LEDs are connected to PORT C of a PIC16F887 microcontroller
The program turns the LEDs ON and OFF alternately, giving the impression that
the LEDs are moving.

The project is built on an EasyPIC 5 development kit, operating with a
8MHz crystal.

```
   Programmer       : Dogan Ibrahim
   File             : PROJECT2.C
   Microcontroller  : PIC16F887
***********************************************************/
```

```
void main()
{
    unsigned char Count = 1;            // Initialize Count to 1
    TRISC = 0;                          // PORT C pins are outputs

    for(;;)                             // Endless loop
    {
        PORTC = Count;                  // Display Count
        Delay_Ms(1000);                 // Wait 1 seconds
        Count = Count << 1;             // Shift Count left
        if(Count == 0)Count = 1;        // If Count=0, set it to 1
    }
}
```

Figure 5.14 Program listing

5.5.6 Suggestions for Future Work

The project can be modified such that the LEDs are shifted left and right alternately with a 1 second interval.

5.6 PROJECT 3 – LED WITH PUSH BUTTON SWITCH

5.6.1 Description

In this project an LED is connected to port pin RC0 of PORT C. In addition, a push-button switch is connected to port pin RC7 of the microcontroller. The LED is normally OFF and starts flashing with 1 second intervals when the button is pressed.

5.6.2 Block Diagram

The block diagram of the project is as shown in Figure 5.15.

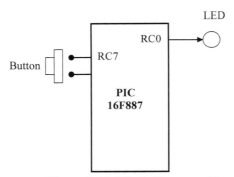

Figure 5.15 Block diagram of the project

5.6.3 Circuit Diagram

The circuit diagram of the project is as shown in Figure 5.16.

5.6.4 PDL of the Project

The PDL of the project is shown in Figure 5.17. The LEDs are flashed continuously with an interval of one second.

        BEGIN
            Configure RC0 as output and RC7 as input
            Turn LED OFF to start with
            Wait until the button is pressed
            DO FOREVER

    Turn LED ON
    Wait 1 second
    Turn LED OFF
    Wait 1 second
  ENDDO
END

Figure 5.17 PDL of the project

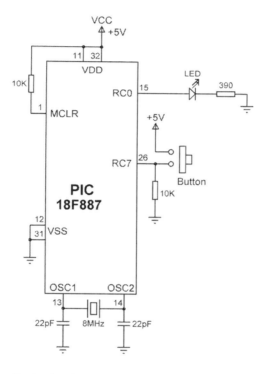

Figure 5.16 Circuit diagram of the project

The following additional jumpers must be selected on the EasyPIC 5 development board (see Figure 5.18):

  PORT C Pull-Down Switch 8 ON (RC7 ON)
  PORT C Pull-Down selected
  J17 VCC (RC7 goes to VCC when button RC7 pressed)

Figure 5.18 EasyPIC 5 jumper settings for the **button**

5.6.5 Program Listing

The program listing of the project (PROJECT3.C) is shown in Figure 5.19. At the beginning of the project PORT C pins are configured such that RC7 (MSB) is an input and RC0 (LSB) is an output. Then the program waits until the button is pressed. i.e. until RC7 pin of the microcontroller becomes logic 1. The program then enters an endless loop while the LED is turned ON and OFF with 1 second interval.

```
/***********************************************************
            BUTTON CONTROLLED LED
            **************************
```

In this project an LED is connected to RC0 pin of a PIC16F887 microcontroller. In addition, push-button switch is connected to RC7 pin. After a RESET the LED is normally OFF. The LED starts flashing after the switch is pressed.

Pin RC7 of the microcontroller is normally at logic 0, and goes to logic 1 when the button is pressed.

The project is built on an EasyPIC 5 development kit, operating with a 8MHz crystal.

```
Programmer       : Dogan Ibrahim
File             : PROJECT3.C
Microcontroller  : PIC16F887
```

```
******************************************************************/
#define Button PORTC.F7

void main()
{
   TRISC = 0x80;                   // RC0 is output, RC7 is input
   PORTC = 0;                      // Turn OFF the LED to start with

   while(Button != 1);             // Wait until button is pressed

   for(;;)                         // Endless loop
   {
       PORTC.F0 = 1;               // Turn LED ON
       Delay_Ms(1000);             // Wait 1 seconds
       PORTC.F0 = 0;               // Turn LED OFF
       Delay_Ms(1000);             // Wait 1 second
   }
}
```

Figure 5.19 Program listing

Note that the statement PORTC.F0 accesses bit 0 of PORT C (i.e. RC0).

5.5.6 Suggestions for Future Work

The project can be modified such that the LEDs can count up in binary when a button is pressed and count down when the button is released.

5.7 PROJECT 4 – COUNTING LCD

5.7.1 Description

In this project an LCD is connected to a PIC16F887 type microcontroller. The LCD counts up with one second intervals.

5.7.2 Block Diagram

The block diagram of the project is as shown in Figure 5.20.

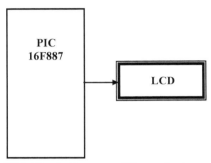

Figure 5.20 Block diagram of the project

In microcontroller systems the output of a measured variable is usually displayed using LEDs, 7-segment displays, or LCD type displays. LCDs have the advantages that they can be used to display alphanumeric or graphical data. Some LCDs have 40 or more character lengths with the capability to display several lines. Some other LCD displays can be used to display graphics images. Some modules offer colour displays while some others incorporate back lighting so that they can be viewed in dimly lit conditions.

There are basically two types of LCDs as far as the interfacing technique is concerned: parallel LCDs, and serial LCDs. Parallel LCDs (e.g. Hitachi HD44780) are connected to a microcontroller using more than one data line and the data is transferred in parallel form. It is common to use either 4 or 8 data lines. Using a 4 wire connection saves I/O pins but it is slower since the data is transferred in two stages. Serial LCDs are connected to the microcontroller using only one data line and data is usually sent to the LCD using the standard asynchronous serial data communication methods. Serial LCDs are much easier to use but they cost more than the parallel ones.

The low level programming of a parallel LCD is usually a complex task and requires a good understanding of the internal operation of the LCD controllers, including the various timing requirements. Fortunately, mikroC language provides special library commands for displaying data on alphanumeric as well as on graphical LCDs. All the user has to do is connect the LCD to the microcontroller, define the LCD connection in the software, and then send special commands to display data on the LCD.

HD44780 LCD Module

HD44780 is one of the most popular alphanumeric LCD modules used in industry and also by hobbyists. This module is monochrome and comes in different sizes. Modules with 8, 16, 20, 24, 32, and 40 columns are available. Depending on the model chosen, the number of rows varies between 1,2 or 4. The display provides a 14-pin (or 16-pin) connector to a microcontroller. Table 5.1 gives the pin configuration and pin functions of a 14-pin LCD module.

Table 5.1 Pin configuration of HD44780 LCD module

Pin no	Name	Function
1	V_{SS}	Ground
2	V_{DD}	+ ve supply
3	V_{EE}	Contrast
4	RS	Register select
5	R/W	Read/write
6	E	Enable
7	D0	Daat bit 0
8	D1	Data bit 1
9	D2	Data bit 2
10	D3	Data bit 3
11	D4	Data bit 4
12	D5	Data bit 5
13	D6	Data bit 6
14	D7	Data bit 7

Below is a summary of the pin functions:

V_{SS} is the 0V supply or ground. The V_{DD} pin should be connected to the positive supply. Although the manufacturers specify a 5V D.C. supply, the modules will usually work with as low as 3V or as high as 6V.

Pin 3 is named V_{EE} and this is the contrast control pin. This pin is used to adjust the contrast of the display and it should be connected to a variable voltage supply. A potentiometer is normally connected between the power supply lines with its wiper arm connected to this pin so that the contrast can be adjusted.

Pin 4 is the Register Select (RS) and when this pin is LOW, data transferred to the display is treated as commands. When RS is HIGH, character data can be transferred to and from the module.

Pin 5 is the Read/Write (R/W) line. This pin is pulled LOW in order to write commands or character data to the LCD module. When this pin is HIGH, character data or status information can be read from the module.

Pin 6 is the Enable (E) pin which is used to initiate the transfer of commands or data between the module and the microcontroller. When writing to the display, data is transferred only on the HIGH to LOW transition of this line. When reading from the display, data becomes available after the LOW to HIGH transition of the enable pin and this data remains valid as long as the enable pin is at logic HIGH.

Pins 7 to 14 are the eight data bus lines (D0 to D7). Data can be transferred between the microcontroller and the LCD module using either a single 8-bit byte, or as two 4-bit nibbles. In the latter case only the upper four data lines (D4 to D7) are used. 4-bit mode has the advantage that four less I/O lines are required to communicate with the LCD. In this book we shall be using alphanumeric based LCD only and look at the 4-bit interface only.

Connecting the LCD

mikroC compiler assumes by default that the LCD is connected to the microcontroller as follows:

LCD	Microcontroller port
D7	Bit 7 of the port
D6	Bit 6 of the port
D5	Bit 5 of the port
D4	Bit 4 of the port
E	Bit 3 of the port
RS	Bit 2 of the port

Where port is the port name specified using the **Lcd_Init** statement. For example, we can use the statement **Lcd_Init(&PORTB)** if the LCD is connected to PORT B with the above default connections.

It is also possible to connect the LCD differently to the microcontroller and use the command **Lcd_Config** to define the connection.

microC compiler supports the following LCD functions (further details can be obtained from the mikroC Manual):

Lcd_Config: This function defines the connections between the LCD and the microcontroller. The arguments should be as in the following order:

        Port name, RS, EN, WE, D7, D6, D5, D4

Lcd_Init: This function initializes the LCD.

Lcd_Out: This function displays text at the given row and column of the LCD.

Lcd_Out_Cp: This function displays text at the current cursor position of the LCD.

Lcd_Chr: This function displays a character at the given row and column position of the LCD.

Lcd_Chr_Cp: This function displays a character at the current row and column position of the LCD.

Lcd_Cmd: This function sends a command to the LCD. Some of the popular commands are:

    LCD_CLEAR: Clear display
    LCD_UNDERLINE_ON: Turn ON underline
    LCD_BLINK_CURSOR_ON: Turn ON blink
    LCD_RETURN_HOME: Return cursor to home position
    LCD_FIRST_ROW: Move to first row
    LCD_SECOND_ROW: Move to second row

5.7.3 Circuit Diagram

In this project. the EasyPIC 5 development board with an LCD is used to develop and test the system. The circuit diagram of the project is as shown in Figure 5.21. The connections between the LCD and the PIC16F887 microcontroller are as follows:

LCD	Microcontroller
D7	Bit 3 of PORT B
D6	Bit 2 of PORT B
D5	Bit 1 of PORT B
D4	Bit 0 of PORT B
E	Bit 5 of PORT B
RS	Bit 4 of PORT B

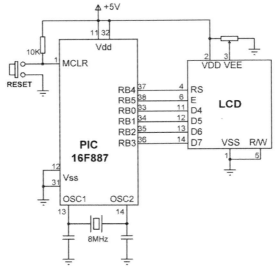

Figure 5.21 Circuit diagram of the project

The following jumper should be set on the EasyPIC 5 development board in order to enable the LCD backlight:

Set: SW9 switch LCD to VCC

5.7.4 PDL of the Project

The PDL of the project is shown in Figure 5.22. The operation of the software is very simple. After configuring and initializing the LCD, a variable counts up by 1 every second and its value is converted to a string and sent to the LCD.

BEGIN
    Configure and initialize the LCD
    Count = 0
    DO FOREVER
        Clear LCD display and Home cursor
        Convert count to a string
        Send Count to the LCD
        Wait 1 second
        Increment Count
    ENDDO
END

Figure 5.22 PDL of the project

5.7.5 Program Listing

The program listing of the project (PROJECT4.C) is shown in Figure 5.23. At the beginning of the program variable **Count** is initialized to 0, PORT B pins are configured as digital I/O, and the LCD connections are defined using built-in library function **Lcd_Config**. Then, an endless loop is formed and inside this loop the value of **Count** is converted into string using built-in function **ByteToStr**. The LCD is then cleared and the cursor is positioned at the Home position. The string value of variable **Count** is displayed on the LCD, the program waits 1 second, and after incrementing the value of **Count** the above process is repeated.

```
/******************************************************************
                    COUNTING LCD
                  ***************
```

In this project an LCD is connected to PORT B of a PIC16F887 type microcontroller. The microcontroller counts up by one (from 0 to 255) with one second interval and the count is displayed on the LCD.

The project is built on an EasyPIC 5 development kit, operating with a 8MHz crystal.

```
Programmer      : Dogan Ibrahim
File            : PROJECT4.C
Microcontroller : PIC16F887
******************************************************************/
```

```
void main()
{
    unsigned char Count = 0;
    char Txt[4];

    TRISB = 0;                          // PORT B is output
    ANSELH = 0;                         // PORT B pins are digital I/O
    Lcd_Config(&PORTB,4,5,7,3,2,1,0);   // Configure LCD connections

    for(;;)                             // Endless loop
    {
        ByteToStr(Count, Txt);          // Convert Count to string
        Lcd_Cmd(LCD_CLEAR);             // Clear LCD
        Lcd_Cmd(LCD_RETURN_HOME);       // Home position
        Lcd_Out_Cp(Txt);                // Display Count
        Delay_Ms(1000);                 // Wait 1 second
        Count++;                        // Increment count
    }
}
```

Figure 5.23 Program listing of the project

Figure 5.24 shows a picture of the EasyPIC 5 board with the LCD display.

Figure 5.24 EasyPIC 5 board with the LCD display

5.7.6 Suggestions for Future Work

The project can be modified such that the project counts and displays external events. The events could be detected as the change in the state of a pin of the microcontroller (e.g. a change of a port pin from 0 to 1, or from 1 to 0).

5.8 PROJECT 5 – SENDING DATA TO A PC USING THE RS232 PORT

5.8.1 Description

In this project a PC is connected to a PIC16F887 type microcontroller. Then the following message is displayed on the PC screen:

Enter a character:

The character entered by the user is read by the microcontroller, incremented by one and then sent back to the PC. For example, if character "A" is entered, the PC displays character "B" and so on.

5.8.2 Block Diagram

The block diagram of the project is as shown in Figure 5.25.

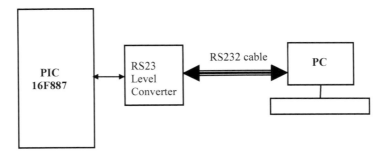

Figure 5.25 Block diagram of the project

The serial communication ports (TXD and RXD) of the microcontroller are connected to a RS232 level converter chip and then to the serial port of a PC (e.g. COM1).

Serial communication is a simple means of sending data to long distances quickly and reliably. The most commonly used serial communication method is based on the RS232 standard. In this standard data is sent over a single line from a transmitting device to a receiving device in bit serial format at a pre-specified speed, also known as the Baud rate, or the number of bits sent each second. Typical Baud rates are 4800, 9600, 19200, 38400 etc.

RS232 serial communication is a form of asynchronous data transmission where data is sent character by character. Each character is preceded with a Start bit, seven or eight data bits, an optional parity bit, and one or more stop bits. The most commonly used format is eight data bits, no parity bit and one stop bit. The least significant data bit is transmitted first, and the most significant bit is transmitted last.

A logic high is defined to be at -12V, and a logic 0 is at +12V. Figure 5.26 shows how character "A" (ASCII binary pattern 0010 0001) is transmitted over a serial line. The line is normally idle at -12V. The start bit is first sent by the line going from high to low. Then eight data bits are sent starting from the least significant bit. Finally the stop bit is sent by raising the line from low to high.

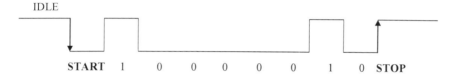

Figure 5.26 Sending character "A" in serial format

In serial connection a minimum of three lines are used for communication: transmit (TX), receive (RX), and ground (GND). Serial devices are connected to each other using two types of connectors: 9-way connector, and 25-way connector. Table 5.2 shows the TX, RX, and GND pins of each types of connectors. The connectors used in RS232 serial communication are shown in Figure 5.27.

Table 5.2 Minimum required pins for serial communication

9-pin connector

Pin	Function
2	Transmit (TX)
3	Receive (RX)
5	Ground (GND)

25-pin connector

Pin	Function
2	Transmit (TX)
3	Receive (RX)
7	Ground (GND)

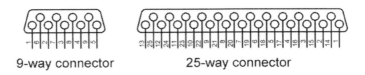

Figure 5.27 RS232 connectors

As described above, RS232 voltage levels are at ±12V. On the other hand, microcontroller input-output ports operate at 0 to +5V voltage levels. It is therefore necessary to translate the voltage levels before a microcontroller can be connected to a RS232 compatible device. Thus, the output signal from the microcontroller has to be converted into ±12V, and the input from an RS232 device must be converted into 0 to +5V before it can be connected to a microcontroller. This voltage translation is normally done using special RS232 voltage converter chips. One such popular chip is the MAX232. This is a dual converter chip having the pin configuration as shown in Figure 5.28. The device requires four external 1μF capacitors for its operation.

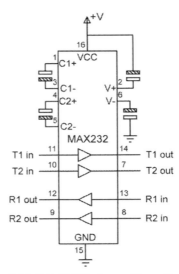

Figure 5.28 MAX232 pin configuration

mikroC compiler supports library functions for both hardware and software type serial communications. Using the hardware option is easy and this is what we will be using in this project. PIC16F887 microcontroller has built-in USART

(Universal Asynchronous Receiver Transmitter) module, providing special input-output pins for serial communication. For serial communication all the data transmission is handled by the USART but we have to configure the USART before receiving and transmitting data. With the software option all the serial bit timing is handled in software and any input-output pin can be programmed and used for serial communication.

The following hardware USART library functions are supported by the mikroC compiler:

Usart_Init: This function initializes the USART module to the specified baud rate.

Usart_Data_Ready: This function is used to test whether or not a character has been received by the USART module.

Usart_Read: This function reads a byte from the USART module.

Usart_Write: This function sends a byte to the USART module.

5.8.3 Circuit Diagram

The circuit diagram of the project is as shown in Figure 5.29. The serial TXD pin (RC6) and serial RXD pin (RC7) of the microcontroller are connected to a MAX232 type RS232 voltage level converter chip. This chip is then connected to a 9-pin D-type connector, compatible with the serial ports of a PC. The PIC16F887 microcontroller is operated from a 8MHz crystal.

5.8.4 PDL of the Project

The PDL of the project is shown in Figure 5.30. The operation of the software is very simple. After configuring and initializing the USART, a character is read from the USART. This character is then incremented by one and sent back to the PC screen.

> **BEGIN**
>   Configure and initialize the USART
>   **DO FOREVER**
>   Display prompt to read a character
>   Read a character from USART
>   Increment the character

      Display the next character on the PC screen
   ENDDO
 END

Figue 5.30 PDL of the project

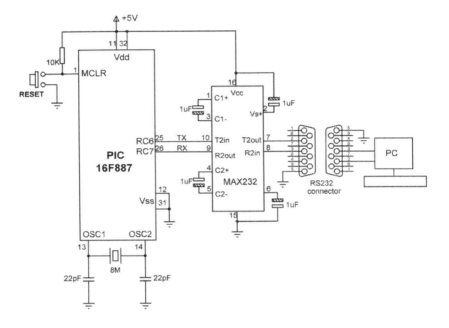

Figure 5.29 Circuit diagram of the project

5.8.5 Program Listing

The program listing of the project (PROJECT5.C) is shown in Figure 5.31. At the beginning of the program USART is initialized to 9600 Baud. Then, the program waits to receive a character from USART. The received character is incremented by one and then sent back to the PC screen. A typical communication with the PC is shown below (the characters entered by the user are in bold for clarity):

 Enter a character: **A**
 The next character is B

Enter a character:
....
....

First of all the program prompts the user to enter a character by displaying the message "Enter a character:" in character array **Msg**, using function **Text_To_USART**. The built-in function **Usart_Data_Ready** is used to detect when a new character is entered by the user. The entered character is stored in variable **MyKey** and echoed to the PC screen. The new character is incremented by one and stored in character array **Result**. This array is then displayed on the PC as "The next character is x" where "x" is the next character in the ASCII table. The program then return to the beginning and the above process is repeated. Notice that a carriage-return and line-feed pair is used to create a newline at the end of each display. In this project the Baud rate is set to 2400.

```
/**********************************************************************
                    SERIAL PC INTERFACE
                    *********************
```

In this project the microcontroller is connected to a PC via its USART port and using a RS232 voltage converter chip.

In this project a character is read from the PC, then this character is incremented by one and the result is displayed on the PC screen.

The project is built on an EasyPIC 5 development kit, operating with a 8MHz crystal.

The Baud rate is set to 2400.

```
    Programmer          : Dogan Ibrahim
    File                : PROJECT5.C
    Micro               : PIC16F887
**********************************************************************/
//
// This functions send carriage-return and line-feed to USART
//
void Newline()
{
   Usart_Write(0x0D);                  // Send carriage-return
   Usart_Write(0x0A);                  // Send line-feed
}

//
// This function sends a text to USART. The text is NULL terminated
//
```

```c
void Text_To_Usart(unsigned char *m)
{
   unsigned char i;

   i = 0;
   while(m[i] != 0)
   {                                         // Send TEXT to USART
     Usart_Write(m[i]);
     i++;
   }
}

//
// Start of MAIN program
//
void main()
{
   unsigned char MyKey, Flag;
   unsigned char Msg[] = "Enter a character: ";
   unsigned char Result[] = "The next character is ";
//
// Configure the USART
//
   Usart_Init(2400);                         // Baud=2400


//
// Start of program loop
//
   for(;;)                                   // Endless loop
   {
     Flag = 1;
     Newline();                              // Send newline
     Text_To_Usart(Msg);                     // Send prompt
//
// Get a character
//
     do
     {
       if(Usart_Data_Ready())                // If a character ready
       {
         MyKey = Usart_Read();               // Get a character
         Usart_Write(MyKey);                 // Echo the character
         Newline();                          // Newline
         MyKey++;                            // Increment the character
         Result[22] = Mykey;                 // New character in output array
         Text_To_Usart(Result);              // Display the result
         flag = 0;
       }
```

```
    }while(flag == 1);
  }
}
```

Figure 5.31 Program listing

5.8.6 Testing the Program

The program can be tested by connecting the 9-pin D-type connector to the serial port (COM1: or COM2:) of a PC. Some new PCs do not have any serial ports and in these cases a USB-to-serial converter device can be used to provide serial port capability to a PC.

If you are using the EasyPIC 5 development board connect the 9-pin connector to the RS232 serial port of the board (see Figure 5.32) and set the following switches on the board (see Figure 5.32):

> Switch SW7 – 1 ON (enable RXD)
> Switch SW8 – 1 ON (enable TXD)

Figure 5.32 EasyPIC 5 board settings

The program can be tested using a terminal emulator software such as the **HyperTerminal** which is distributed free of charge with the Windows operating systems. The steps to test the program are given below (these steps assume that serial port COM2 is used):

- Connect the RS232 output from the microcontroller to the serial input port of a PC (e.g. COM2).

- Start HyperTerminal terminal emulation software and give a name (e.g. TEST) to the session (see Figure 5.33). Click OK

Figure 5.33 The HyperTerminal

- Select the serial port (e.g. COM2) as shown in Figure 5.34. If you are not sure which serial ports are available on your PC, run **Start -> Settings -> Control Panel -> System -> Hardware -> Device Manager -> Ports** to see the available serial port names.

- Select the serial port parameters as (see Figure 5.35):

    Baud: 2400
    Date: 8 bits
    Parity: None
    Stop bits: 1
    Handshaking: None

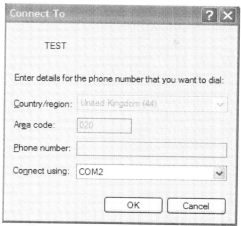

Figure 5.34 Select serial port

Figure 5.35 Set serial port parameters

- Reset the microcontroller

An example output from the HyperTerminal screen is shown in Figure 5.36.

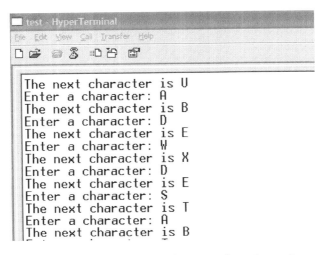

Figure 5.36 An example output from the project

5.8.7 Suggestions for Future Work

A simple four-function integer calculator program can easily be developed. Read two numbers and the function to be performed (add. subtract, multiply, divide) from the PC keyboard. Then, calculate and display the result on the PC screen.

CHAPTER 6

THE GSM/GPRS MODEM

6.1 Overview

This chapter describes the basic operating and interfacing principles of a GSM/GPRS modem that can be used in microcontroller applications. The **Smart GSM/GPRS Development Board** (manufactured by mikroElektronika) is used in the Modem projects in this book.

Before looking at the details of the GSM/GPRS development board and its use in projects, it is worthwhile to review the basic principles of the GSM system.

6.2 The GSM System

GSM (*Global System for Mobile Communications*) is currently the most widely used standard for mobile telephone systems in the world. It is estimated that over 80% of the mobile market uses the GSM standard, with estimated 3 billion people across over 200 countries.

GSM is a cell based (cellular) network where mobile phones access the network by connecting to the cell in their immediate vicinities. GSM is based on an all digital technology. GSM networks operate in a number of different carrier frequencies. In Europe most networks operate at either 900 or 1800MHz. In most of Americas the carrier frequency is either 850 or 1900MHz. In the UK there are over 75 million subscribers. Table 6.1 shows the main GSM operators in the UK with their frequency bands. Other GSM operators (e.g. Virgin Mobile, Asda Mobile, Tesco Mobile etc. use services of the main operators). Some mobile phones are designed to be quad-band and they cover the frequencies 800/900/1800/1900MHz. Such phones can be used in most countries. Triband phones cover the frequency bands 900/1800/1900MHz and can be used in most of Europe and the Americas. The transmission power in the handset is limited to 2 Watts in the 800/900MHz bands, and to 1 Watt in the 1800/1900MHz bands. The new 3G standard phones use the frequency band 2100MHz. The GSM system was designed with a moderate level of security where the subscriber is authenticated using a pre-shared key, and the communication between the subscriber and the network is encrypted.

Table 6.1 GSM operators in the UK

GSM operator	Carrier Frequency (MHz)	Ownership
Orange/T-Mobile	1800/2100	France Telecom
O_2	900/1800/2100	Telefonica
Vodafone	900/1800/2100	Vodafone

GSM networks are usually divided into three segments, these being the:

1. **Radio Access Network**

This segment of the network deals with communication between terminals (handsets, etc.) and Base Transceiver Stations (BTS) (known as 'node B' in 3G networks). The LAPDm protocol is used for signalling, authentication, and the like. BTS stations can vary in size and capacity, from serving several hundreds of users to several thousand, and serving a radius from a few hundred metres to several kilometres.

2. **Base Station Subsystem (BSS)**

This subsystem consists of Base Transceiver Stations and Base Station Controllers (BSC). It provides a link between the mobile terminal and Network Switching Subsystem (NSS). Typically, several BTS are controlled by a single BSC, though for resilience and capacity the configuration is sometimes modified.

3. **Network Subsystem (NSS)**

The network subsystem provides the mechanisms (via the use of several 'registers') for payment and billing, identification and location, and also provides further outbound routing to PSTN telephone networks and IP networks.
The Mobile Switching Centre (MSC), within the NSS, performs the necessary conversion/de-modulation from digital to analogue and vice-versa.

Since GSM is a mobile network, it is inevitable that users will leave the coverage of one BTS and move to another. In order that communications are not dropped, a technique known as 'handover' or 'handoff' is employed to provide seamless transition.

There are several types of handover, depending on whether the user is leaving the area covered by a BSC, or further still, the area covered by an MSC.

The two main types of handover are as follows:

1. Hard Handover

This is where a connection between a terminal and BTS is broken before another is established. The handoff is performed instantly, and in most cases the user is not aware of the handoff. However, in certain scenarios, such as when travelling in a vehicle at high-speed, the handoff procedure may inadvertently drop the connection, if for example, the user has strayed into a different area covered by a different BTS, while handoff is still taking place.

2. Soft Handover

Soft handover involves making a new connection, to the target BTS, before the existing connection is broken. For a short duration, both BTS will be connected to the same handset. This, of course, makes it a far more seamless transition. This method also allows the handset to use the BTS with the best signal quality, until handover is complete.

Handover is, of course, dependent on resources available at the BTS. If another channel cannot be utilised by a BTS (for example, when there is high channel utilisation at the time of handover), then the connection will be dropped, and the user would have to redial. If the utilisation is still high, and there no free channels, the user will be unable to make the call until resources are again available (e.g. until another user ends their connection.)

6.3 The SIM Card

The operation of a GSM/GPRS modem (or a mobile phone) require a SIM card. The SIM (Subscriber Identity Module) card is a small (15mmx25mmx0.76mm) smart card used to store the subscriber information and identify of a subscriber to the cellular network. A user can change handsets by simply removing the SIM card from one phone and inserting it into another phone. The use of a SIM card is mandatory in a mobile phone as it stores the International Mobile Subscriber

Identity (IMSI) number, security authentication and ciphering information, password (card pin number), phone book details, log of sent and received messages, received and dialled call log, and so on. Figure 6.1 shows the layout of a typical SIM card used in mobile phones. The card has six pins and is usually held in a SIM card holder (see Figure 6.2). Card pins have the following definitions (as specified by the ISO 7816 standard):

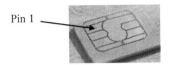

Figure 6.1 SIM card layout

Pin 1 (VCC): This pin provides the card's power supply. A SIM card is typically operated between 3V and 5V.

Pin 2 (RST): This is the RESET pin. A low-level on this pin resets the card to a known state.

Pin 3 (CLK): This pin provides the card with the clock signals. The clock is between 1 and 5 MHz, with a duty cycle between 40% and 60%.

Pin 4 (GND): This is the ground pin of card's power supply.

Pin 5 (VPP): This pin is not used anymore. It was used to provide programming voltage required to write and erase the card's internal non-volatile memory.

Pin 6 (I/O): This is the input and output pin of the card. Half-duplex asynchronous serial communication is used to read and write data to the card.

When the card contacts are activated the RST pin is held at a low level and the VCC supply is applied. After maintaining the RST signal at a low state for at least 40,000 clock cycles it is raised to a high state and communication with the card starts by driving the I/O line while the RST is held high.

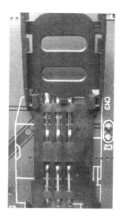

Figure 6.2 A typical SIM card holder

6.4 The Smart GSM/GPRS Development Board

The Smart GSM/GPRS board (see Figure 6.3) is developed and manufactured by mikroElektronika (www.mikroe.com). The board has the following specifications:

- Support for 5 different GSM modems
- SIM card holder
- Antenna holder
- Audio amplifier and audio interface with microphone and speaker
- RS232 interface for PC or microcontroller interface
- LED showing the modem status
- Interface signals on 10-way connectors for easy interface to microcontroller development systems

The board is powered by using an external 12V power supply. Sockets are provided on the main board so that various GSM/GPRS modem cards (daughter boards) can be plugged onto the main Smart GSM/GPRS development board. The operation of a GSM/GPRS modem requires a SIM card (used in mobile phones) and a card holder is available on the main board to hold a SIM card. The main board can be interfaced directly to a PC using its RS232 port. Thus, the main board and the modem card can be controlled from a PC. Either a program can be developed on the PC, or a terminal emulation program (e.g. HyperTerminal) can be used for this purpose. A microphone, speaker, and an audio amplifier are available on the main board so that speech communication can be established using the GSM/GPRS modem. The modem status is indicated by a small LED mounted on the main board. The LED flashes at different rates depending on whether or not the system is connected to a GSM/GPRS base station. All of the interface signals are available on 10-way connectors at the edges of the main board. Thus, the main board can be connected to an external system (e.g. to a microcontroller) via these connectors for its control and operation.

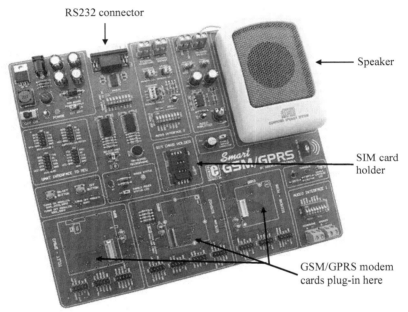

Figure 6.3 The Smart GSM/GPRS Development Board

In this book the SIM340Z GSM/GPRS modem (manufactured by SIMCOM) is used. This is a quad-band GSM/GPRS modem with the following features:

- 850/900/1800/1900 MHz operation
- Small footprint (40mmx33mmx2.85mm)
- Low weight (8g)
- GSM 07.07 and 07.05 compatible
- Control via AT commands
- Low power operation (3.4V – 4.5V)
- SIM card interface
- Keypad and LCD interfaces

The SIM340Z modem[7] (see Figure 6.4) is easily interfaced to a microcontroller using only three serial communication pins (TXD, RXD, and RTS). The modem provides signals to drive a SIM card directly. In addition, an LED can be connected to the modem to show the working status of the modem (see Table 6.2). The modem also provides a buzzer interface where a buzzer or a small speaker can be connected to indicate an incoming call. In addition, the modem provides signals for a keypad and an LCD to be connected for dedicated autonomous mobile communication applications. A small connector is provided on the modem for an external antenna to be connected.

Table 6.2 LED showing the modem status

LED State	SIM340Z Function
OFF	Modem is OFF
64ms ON/800ms OFF	GSM network not found
64ms ON/3000ms OFF	GSM network found
64ms ON/300ms OFF	GPRS communication

Figure 6.5 shows a functional block diagram of the SIM340Z modem card. The modem card provides all the signals required in a typical basic mobile phone. The card provides signals to drive a SIM card, microphone input, speaker output, a display, a keypad interface, status LED, and buzzer output. A small buzzer can be connected to the buzzer output to indicate the ringing tone. The microphone and speaker pins are used for speech applications.

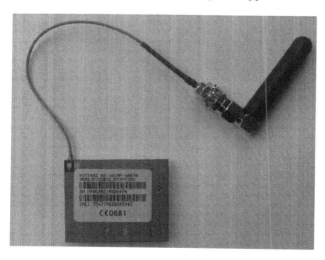

Figure 6.4 SIM340Z GSM/GPRS modem card

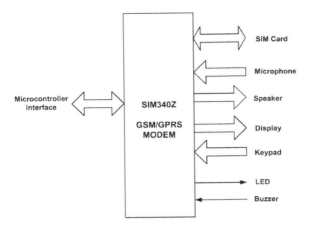

Figure 6.5 SIM340Z functional block diagram

6.5 Using the Smart GSM/GPRS Board With a PC

In this section we shall be looking at how the Smart GSM/GPRS board can be used directly with a PC. Several projects will be given to show how the board can be used in speech and SMS based communications applications.

6.5.1 Setting-Up the Hardware

Before using the Smart GSM/GPRS board with a PC you should do the following (see Figure 6.6):

- Insert a valid SIM card into its holder on the board
- Set jumpers labelled "ENABLE USART INTERFACE TO PC" to ON position. This switch enables the RS232 level converter chips
- Set jumper J1 (SELECT UART INTERFACE) to position PC
- Set jumper J15 (ENABLE POWER STATUS LED)
- Set jumper J2 (ENABLE TIMEPULSE LED)
- Connect the antenna of the SIM340Z modem card and plug-in the card onto the main board
- Connect a RS232 cable from the main board (RS232 connector CN7) to the PC (if the PC does not have any RS232 ports you can use a USB-to-RS232 converter device to provide RS232 interface to your PC)
- Connect an external power supply (e.g. 12V) to the main board

- Turn ON the power supply and the POWER SUPPLY switch

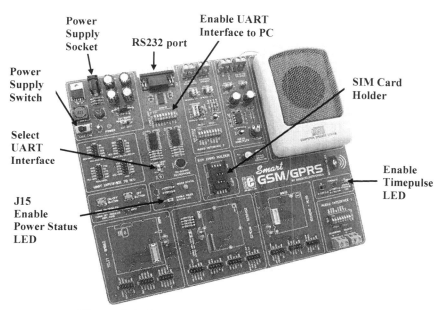

Figure 6.6 Smart GSM/GPRS jumper and switch locations

After turning ON the power supply, the POWER STATUS LED should turn ON and stay ON continuously. The TIMEPULSE LED should flash with short intervals to indicate that the board is ready and can communicate with a local GSM network.

Figure 6.7 shows the circuit diagram of the hardware setup for the readers who do not have a Smart GSM/GPRS board, and who may want to design the hardware on a PCB or a breadboard. One of the important points to remember here is that the SIM340Z modem card operates with a power supply voltage between 3.4V – 4.5V. In this example a 4.0V power supply is used (a 5V regulator can be used to generate power for the MAX232 chip. Then, a regulator can be used to obtain the 4.0V required for the modem card).

It is also important to realise that the input-output pins of the modem card are not compatible with a 5V CMOS chip such as the MAX232. The maximum input voltage at any pin of the modem card is about 3.0V. The maximum logic HIGH output voltage of the MAX232 (e.g. pin Tio) is about 5V and this is too large an input for the modem input pin (e.g RXD). Thus, the output of the MAX232 can not directly be connected to the modem card. Similarly, the maximum logic high output voltage of a pin of the modem card (e.g the TXD pin) is about 2V and this is not enough to drive the logic input of the MAX232 (e.g pin R1i). Thus, voltage level converters are required between the MAX232 chip and the modem card. The following pins of the modem card are used to interface it to the serial port:

- TXD
- RXD
- RTS
- CTS
- DTR
- DCD
- RI

On the Smart GSM/GPRS board, SN74LVCC3245 type octal voltage level translator chips are used to provide the correct signals between the MAX232 and the modem card. This chip has dual power supply inputs (3V and 5V) and thus enables a 3V logic chip to be connected to a 5V logic chip. VCCA power input accepts 2.3V to 3.6V (connected to the VDD_EXT output of the modem card which provides 2.93V output), and the VCCB power input accepts voltages 3V to 5.5V (connected to normal 5V power supply). Pin DIR controls the signal direction. When DIR is at logic 0, "B" data is transferred to "A", and when DIR is at logic 1, "A" data is transferred to "B". Input OE controls the 3-state outputs of the chip and this input should normally be at logic 0 for normal operation.

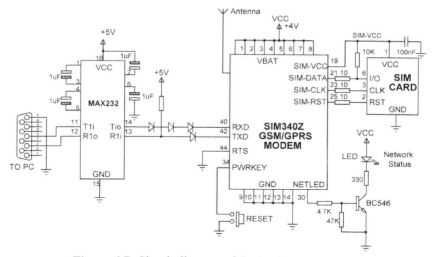

Figure 6.7 Circuit diagram of the hardware setup

Although 7 pins are used for serial communication, in normal situations when communicating with a PC only the pins RXD, TXD and GND are sufficient. In addition, the RTS pin should be connected to logic 0. In such cases where only a few pins are required, use of level translator chips can be expensive and as shown in Figure 6.7, it might be cheaper to use resistive potential divider circuits or diode-resistor circuits to provide the required voltage translation between a 3V logic and a 5V logic.

In Figure 6.7, three semiconductor diodes are used to lower the 5V output of MAX232 by 2.1V to about 3V for the RXD input. Similarly, a semiconductor diode and a resistor are used to increase the TXD output of the modem card by about 0.7V (which is the voltage drop across the diode) to drive the R1i input of the MAX232.

Notice that the SIM card is powered and also driven directly from the modem card.

6.5.2 Setting-Up the PC Software

After setting-up the hardware we should use a terminal emulation software on our PC so that communication can be established with the PC. Assuming that the **HyperTerminal** software is being used the steps are as given in section 5.7.6, except that the Baud rate should be set to 19200.

Press the RESET key of the modem and keep it down for at least 3 seconds. The Timepulse LED (Network Status LED) will start flashing. Then, press the ENTER key on the PC keyboard and then enter command **AT** (in upper case and followed by ENTER. Note that **At** or **aT** or **at** are not allowed while auto detecting the baud rate. The commands are however case insensitive after the baud rate has been detected). The modem will auto-detect the PC baud rate and set its baud rate also to 19200, and as shown in Figure 6.8 it will respond with **OK**. We are now ready to communicate with the modem by sending commands or monitoring the state of the modem using the **AT** commands.

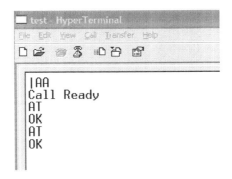

Figure 6.8 Modem responding with OK

6.5.3 AT Commands

The AT commands are used to control the operations of modems. These commands were first developed for the Hayes Smartmodem 300 in late 1970s. An AT command consists of the letters "**AT**" followed by a number of characters specifying the command tail. Some commands are used to set the configuration of modems, some are used to interrogate modems and get their configurations, while some other commands are used to dial numbers, send SMS messages and so on.

In addition to the standard **AT** command set, the SIM340Z GSM/GPRS modem supports commands to configure the modem and send SMS messages to mobile phones, and to send FTP files (*further information can be obtained from the SIMCOM Manuals and Application Notes*).

An AT command can be issued for the following reasons:

- To check the modem settings. e.g. **AT+param=?** where the modem lists the allowable values of the chosen parameter **param**
- To check the current value of a setting. e.g. **AT+param?** displays the setting of parameter **param**
- To change a modem setting. e.g. **AT+param=x** changes value of setting **param** to **x**
- To issue a command. e.g. **AT+D** sends command D to the modem

Table 6.3 lists some of the commonly used general purpose AT commands. Some examples are given in this section to familiarize the user with the use of these commands.

Table 6.3 Some general purpose AT commands

AT command	Desciption
ATI	Modem product information
ATE0	Disable echo
ATE1	Enable echo
ATS3	Command termination character
AT+GMI	Modem manufacturer's id
AT+GMM	Modem model number

Command **ATI** displays the modem product information as shown in Figure 6.9.

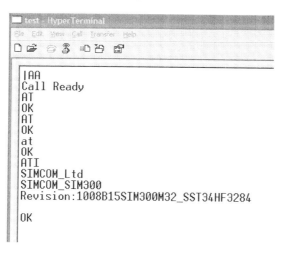

Figure 6.9 Command ATI displays modem product information

Command **ATE0** disables the Echo mode so that characters typed by the user are not echoed back by the modem. Similarly, command **ATE1** enables the Echo mode. In Figure 6.10 echo mode is disabled and command **ATI** is entered. Notice that user typed command **ATI** is not displayed by the modem. Then the echo mode is enabled and command **ATI** is re-entered.

```
OK
ATE0
OK

SIMCOM_Ltd
SIMCOM_SIM300
Revision:1008B15SIM300M32_SST34HF3284

OK

OK
ATI
SIMCOM_Ltd
SIMCOM_SIM300
Revision:1008B15SIM300M32_SST34HF3284

OK
```

Figure 6.10 Enabling and disabling the echo mode

Command ATS3 is used to change the command termination character. By default this is the Linefeed character (decimal 13). As shown in Figure 6.11, command ATS3? can be used to display the current setting of this parameter.

Figure 6.11 Displaying the current value of command termination character

126

Command **AT+GMI** displays the modem manufacturer's identification (see Figure 6.12).

Command **AT+GMM** displays the modem model number (see Figure 6.13).

Figure 6.12 Displaying modem manufacturer's identification

Figure 6.13 Displaying the modem model number

6.6 GSM/GPRS – PC PROJECT 1 – SENDING SMS MESSAGES

In this section we shall see how an SMS message can be sent to a mobile phone using our Smart GSM/GPRS board. Before going into details of the project it is worthwhile to see what an SMS is and what commands are available for sending an SMS to a mobile phone.

In addition to the standard AT command set, the SIM340Z GSM/GPRS modem supports commands to configure the modem and send SMS messages to mobile phones. Table 6.4 gives a list of the important AT commands available *for sending* an SMS message (*further information can be obtained from the SIMCOM SMS Application note AN_SMS_V1.01*).

Table 6.4 Important SIM340Z modem SMS commands

AT Command	SMS Function
AT+CMGF	Select SMS message format
AT+CMCS	Select SMS character set
AT+CSCA	SMS service center address
AT+CMGS	Send SMS message
AT+CSMP	Set SMS text mode parameters
AT+CMGD	Delete SMS message
AT+CMGDA	Delete all SMS messages
AT+CMGR	Read SMS message
AT+CMSS	Send SMS message from store
AT+CPMS	Preferred SMS storage
AT+CMGL	List SMS messages from store
AT+CPIN	Enter PIN number
AT+CMGW	Store a message

There are two ways of sending and receiving SMS messages using AT commands: by **PDU** (Protocol Description Unit) mode, and **TEXT** mode. The PDU mode offers to send binary data in 7-bit or 8-bit mode and is helpful if we wish to send compressed data, or binary data. PDU mode data consists of hexadecimal string of characters, including the SMS service center, sender number, time stamp etc.

In this project we shall be sending messages using the easier Text mode. An SMS text message can consist of alphanumeric characters with up to 160 characters long with 7-bit coding, and 140 characters long with 8-bit coding. Basically, SMS is a store-and-forward type service where the messages are not sent directly from the sender to recipient, but via an SMS Service Center (SMSC). Mobile telephone networks have messaging centers that handle the delivery of messages to their destinations. There is by default no confirmation of the delivery of a message. But users can turn ON this option so that a confirmation of delivery can be sent to the sender when a message is delivered successfully.

Sending an SMS message in Text mode is very easy and an example is given below that shows how the message "**My SMS message!**" can be sent to mobile number "**07515932222**" (assuming that the SMS service center number has already been loaded to the SIM card, which is the case by default, and that there is no PIN number associated with the card):

- Set SMS mode to Text:

    AT+CMGF=1

- Set the character mode to GSM:

    AT+CSCS="GSM"

- Set the SMS parameters:

    AT+CSMP=17,167,0,241

- Set the recipient mobile phone number and the text message:

    AT+CMGS="07515932222"
    > My SMS message! [Cntrl-Z]

Note that after sending the recipient mobile phone number, the modem responds with character "**>**" and that the message must be terminated with the "**Cntrl-Z**" character.

The above commands are all that is required to send SMS messages to a mobile phone. Figure 6.14 shows these command entered via the HyperTerminal software.

It is worthwhile to look at briefly the meaning of the various SMS parameters set by command **CSMP**. The format of this command is (further information can be obtained from the AT command interface document):

$\quad$ AT+CSMP = <fo>, <vp>, <pid>, <dcs>
where,

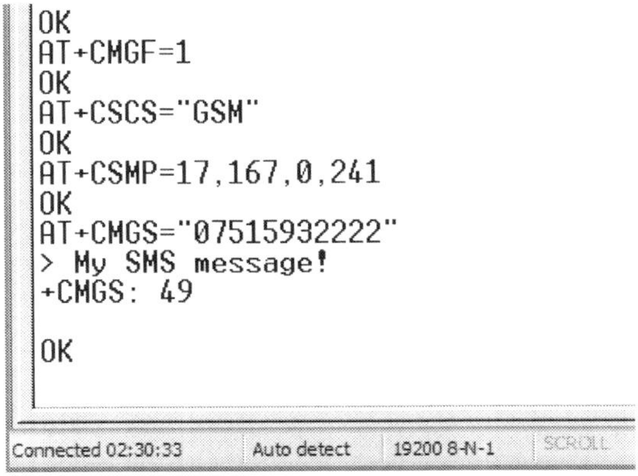
OK
AT+CMGF=1
OK
AT+CSCS="GSM"
OK
AT+CSMP=17,167,0,241
OK
AT+CMGS="07515932222"
> My SMS message!
+CMGS: 49
OK
Connected 02:30:33
Auto detect
19200 8-N-1
SCROLL

Figure 6.14 Sending an SMS message

Field <fo> is set to 17 which indicates that the message is to go from a mobile device to a service center and the <vp> field is valid.

Field <vp> selects the message validity period and a value between 144 to 167 selects the period as:

12 hrs + ((vp – 143) x 30 mins)

With a setting of 167, the message validity period is:

12 hrs + (167 – 143) x 30 mins) = 24 hrs

Field <pid> is used to indicate the higher layer protocol being used and is set to 0 here.

Field <dcs> is used to determine the way the information is encoded. This field is set to 241 which sets the message class to 1 and uses the default alphabet. Note that, setting the message class to 0 causes the message to be displayed immediately without being stored by the mobile phone (this is also called a *flash* message).

In the above example it is assumed that the SMS service center number has already been programmed to the SIM card (which is normally the case when a new SIM card is purchased). The command **AT+CSCA** can however be used to set the SMS service center number.

In addition in the above example it is assumed that the SIM card had no security PIN number associated with it. If the SIM card has a PIN number then command **AT+CPIN** can be used to enter the PIN number.

The command AT+CMGL can be used to list all messages stored on the SIM card. As shown in Figure 6.15, entering command **AT+CMGL=?** lists all the parameters of this command.

```
OK
AT+CMGL=?
+CMGL: ("REC READ","REC UNREAD","STO SENT","STO UNSENT","ALL")
OK
```

Figure 6.15 Listing parameters of command AT+CMGL

We can list all the received SMS messages on the SIM card by entering the command **AT+CMGL="REC READ"** as shown in Figure 6.16.

```
OK
AT+CMGL="REC READ"
+CMGL: 1,"REC READ","+447973100610",,"09/01/12,15:30:27+00"
Hi from Orange. As a thank you for joining we have popped 1 pound onto your acco
unt. This will expire in 1 month. Enjoy!
+CMGL: 2,"REC READ","+447528885664",,"09/01/12,16:28:33+00"
SM1:07528885664,
```

Figure 6.16 Listing all the received SMS messages on the card

Figure 6.16 shows that there are two messages (records) on the SIM card.

We can display any unread messages by entering the command **AT+CMGL="REC UNREAD"**.

All the SMS messages (received and stored and sent) can be displayed by entering the command **AT+CMGL="ALL"** as shown in Figure 6.17 (note that because there were no stored and sent messages, this display is same as Figure 6.16).

```
OK
AT+CMGL="ALL"
+CMGL: 1,"REC READ","+447973100610",,"09/01/12,15:30:27+00"
Hi from Orange. As a thank you for joining we have popped 1 pound onto your acco
unt. This will expire in 1 month. Enjoy!
+CMGL: 2,"REC READ","+447528885664",,"09/01/12,16:28:33+00"
SM1:07528885664,
```

Figure 6.17 Displaying all SMS messages

The command **AT+CMGR=no** can be used to display a single message. For example, as shown in Figure 6.18, entering command **AT+CMGR=1** display no message 1.

```
OK
AT+CMGR=1
+CMGR: "REC READ","+447973100610",,"09/01/12,15:30:27+00"
Hi from Orange. As a thank you for joining we have popped 1 pound onto your acco
unt. This will expire in 1 month. Enjoy!
OK
```

Figure 6.18 Displaying message no 1

The command **AT+CSCA?** Can be used to display the SMS service center number as shown in Figure 6.19 (the number is 0044 7973100973 in this example).

```
OK
AT+CSCA?
+CSCA: "+447973100973",145
OK
```

Figure 6.19 Displaying the SMS service center number

The command **AT+CMGD=no** can be used to delete a single SMS message (**no** is the message number). Figure 6.20 shows how message having number 2 can be deleted from the SIM card. After deleting this message, entering command **AT+CMGL="ALL"** displays only the message with number 1 as this is the only message left on the SIM card.

```
OK
AT+CMGD=2
OK
AT+CMGL="ALL"
+CMGL: 1,"REC READ","+447973100610",,"09/01/12,15:30:27+00"
Hi from Orange. As a thank you for joining we have popped 1 pound onto your acco
unt. This will expire in 1 month. Enjoy!
OK
```

Figure 6.20 Deleting message having identity number 1

Message **AT+CMGDA** can be used to delete all SMS messages on the card.

The command **AT+CMGW** can be used to store a message on the SIM card. Figure 6.21 shows how the message "MY COMPUTER" can be stored in the SIM card:

```
OK
AT+CMGW
> MY COMPUTER
+CMGW: 5

OK
```

Figure 6.21 Storing a message on the SIM card

In the above example the message is stored in memory number 5. We can now send this message to a mobile phone using command **AT+CMSS** as shown in Figure 6.22 (The mobile phone number in this example is "07515932222").

```
OK
AT+CMSS=5,"07515932222"
+CMSS: 47

OK
```

Figure 6.22 Sending message with id no 5 to a mobile phone

It is also possible to specify the mobile phone number with the **AT+CMGW** command and then use the **AT+CMSS** command to send the message. The first command stores the message with a message number and this message number is used in the second command to send the message to the previously specified number. An example is given in Figure 6.23 where the message "Olympics" is first saved in memory number 6, and then this message is sent to mobile phone number "07515932222".

```
...
OK
AT+CMGW="07515932222"
> Olympics
+CMGW: 6

OK
AT+CMSS=6
+CMSS: 48

OK
```

Figure 6.23 Sending a message using **CMGW** and **CMSS**

6.7 GSM/GPRS – PC PROJECT 2 – SENDING SMS MESSAGES IN PDU MODE

In this section we shall be looking at how to send an SMS message in PDU mode. The PDU mode offers to send binary data in 7-bit or 8-bit format. This may be useful if we want to send binary data, or to build our own encoding of the characters.

In general, SMS text messages can be up to 160 characters long, where each character is 7-bits and the default 7-bit alphabet is used. 8-bit messages can be up to 140 characters long, but they can not be used in mobile phones as text messages, instead they are used to send binary data. 16-bit messages can be up to 70 characters long and are used for UCS2 (Unicode) text messages, only viewable by some mobile phones.

PDU mode is more difficult than the text mode SMS for the user interface. The PDU mode contains not only the message, but also a lot of additional data such as information about the sender, the SMS service center number, time stamp, etc. All the characters used in the PDU mode are in hexadecimal number format, called octets, or decimal semi-octets.

Sending data in the PDU mode is similar to sending in SMS mode. The steps are as follows:

- Set SMS mode to PDU:

    AT+CMGF=0

- Specify number of characters to send (excluding two initial zeros):

    AT+CMGS=19
    >0011000B913201183451F60000AA05E8329BFD06 [Cntrl-Z]

There are 19 octets (38 characters) in this message (excluding the first octet ("00"). Note that after specifying the number of characters, the modem responds with character "**>**" and that the message must be entered as hexadecimal numbers, and must be terminated with the **"Cntrl-Z"** character.

The above commands are all that is required to send SMS messages in PDU mode to a mobile phone. The various fields of a PDU type message are described below.

The format of a PDU message is as shown in Figure 6.24.

1-12 octet	1 octet	1 octet	1 octet	1 octet	2-12 octet	1 octet	1 octet	1 octet	1 octet	0-140 octet
SCA	PDU type	MR	LEN	TNO	DA	PID	DCS	VP	UDL	UD

Figure 6.24 Format of a PDU message

Considering the PDU example above where the sent message is:

0011000B913201183451F60000AA05E8329BFD06 [Cntrl-Z]

The fields have the following definitions:

Octets:

00	**SCA** (Service Center Number). Length 0 means that the SMSC is stored in the SIM card.
11	**PDU type** = 11 (binary "0001 0001") to indicate that this is a **Submit Message**. This field does not need to be changed if sending a message
00	**MR**. Message reference number which is incremented from 0 to 255. This field should be set to "00" and is managed by the mobile phone.
0B	**LEN**. Length of the destination phone number (11 decimal here).
91	**TNO**. Type of number. 91 identifies international phone number format. 81 identifies national phone number
3201183451F6	**DA**. The destination phone number in semi-octet form. In this example, the actual number is "23108143156". The length of the phone number is 11 characters long (odd), and therefore a trailing "F" is added to make it even so that it can be shown with semi-octets. If the number is a national number, e.g. 1234567, then the DA field should be "07", PID field should be "81", and the phone number should be "214365F7"
00	**PID**. Protocol Identifier. This field should be "00" for short

	message,
00	**DCS**. Data Coding Scheme. The default "00" indicates that A default 7-bit alphabet is used.
AA	**VP**. Validity Period. "AA" means 4 days. The Validity Period is determined as follows:

        0-143   (VP+1) x 5 minutes
        144-167 12 hours + ((VP-143)x30 mins)
        168-196 (VP-166) x 1 day
        197-255 (VP-192) x 1 week

Hexadecimal "AA" = 170. Thus,

(170-166) x 1 day = 4 days

05	**UDL**. User Message Length (5 in this example)
E8329BFD06	**UD**. User message ("hello" in this example). The user message is encoded as described below

Message Encoding

The message encoding is described in detail in Document GSM 03.38 of the European Telecommunications Standards Institute. The process is described below briefly.

A character consist of 7-bits (a,b,c,d,e,f,g), as:

    b7 b6 b5 b4 b3 b2 b1
    a b c d e f g

If there is only one character ("a"), the top bit is made "0" as in:

    7 6   5   4   3   2   1   0
    0 1a 1b 1c 1d 1e 1f 1g

If there are two characters ("a" and "b"), then two octets are created with leading two zeroes as follows:

	7	6	5	4	3	2	1	0
	2g	1a	1b	1c	1d	1e	1f	1g
	0	0	2a	2b	2c	2d	2e	2f

If there are three characters ("a", "b" and "c"), then three octets are created with leading zeroes as follows:

	7	6	5	4	3	2	1	0
	2g	1a	1b	1c	1d	1e	1f	1g
	3f	3g	2a	2b	2c	2d	2e	2f
	0	0	0	3a	3b	3c	3d	3e

And so on, the bits of the new characters are shifted in. An example is given below to show how the characters "hello" can be decoded for use in SMS PDU mode:

Character	**7-bit pattern**	**Hex code**	**Encoded code**
h	110 1000	68	
e	110 0101	65	
l	110 1100	6C	
l	110 1100	6C	
o	110 1111	6F	

We can start the decoding process:

```
h ->        110 1000

he ->       1 110 1000       insert MSB of "e"
            00 110010        two leading "00"

hel ->      1 110 1000       insert MSB of "e"
            00 110010        two leading "00"
            00 110010        insert 3f 3g
            000 11011        insert 3a 3b 3c 3d 3e

hell ->     1 110 1000       insert MSB of "e"
            00 110010        two leading "00"
            00 110010        insert 3f 3g
            000 11011        insert 3a 3b 3c 3d 3e
            100 11011
            0000 1101        insert 4a 4b 4c 4d
```

hello ->	1 110 1000	insert MSB of "e"	**E8**
	00 110010	two leading "00"	**32**
	000 11011	insert 3a 3b 3c 3d 3e	**9B**
	1111 1101	insert 5d 5e 5f 5g	**FD**
	00000 110	insert 5a 5b 5c	**06**

The encoded octets are: E8329BFD06

An example is given here to show how the SMS message "hello" can be sent in PDU mode to mobile phone number "07515932222". Figure 6.25 shows the commands entered on the PC keyboard.

The PDU data to be sent is:

00 11 00 0B 81 7015952322F2 00 00 AA 05 E8329BFD06

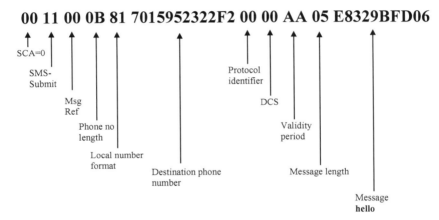

- Set SMS mode to PDU:
    AT+CMGF=0

- Specify number of characters to send (excluding two initial zeros):
    AT+CMGS=19
    > 0011000B817015952322F20000AA05E8329BFD06

Where the PDU message consists of 19 octets (38 characters).

```
OK
AT+CMGF=0
OK
AT+CMGS=19
> 0011000B817015952322F20000AA05E8329BFD06
+CMGS: 50

OK
```

Figure 6.25 Sending SMS message in PDU mode

6.8 GSM/GPRS – PC PROJECT 3 – CALLING A MOBILE PHONE

This section describes a project where a mobile phone number is called and voice communication is established with the mobile phone.

The SIM340 modem includes two microphones and two speaker connections. Smart GSM/GPSR board contains an on board speaker and an on board microphone that can be used for voice communication with a mobile phone. In addition, an external speaker and an external microphone can be connected to the Smart GSM/GPRS board.

The Smart GSM/GPRS board contains audio amplifiers for the speaker and the microphone inputs. Although it is not essential to use audio amplifiers with the modem card, the signal levels can be increased by using audio amplifiers.

To call a mobile phone you should make the following jumper settings on the Smart GSM/GPRS board:

Switch "GSM MODULE SELECTION": Set switches GSM2 (MIC2+) and GSM2 (MIC2-) to ON position

Make sure the on board speaker is enabled (J4 fitted)
Select on board microphone (J9 and J10)
Select on board speaker (J7 and J8)
Select microphone biasing if required (J11,J12,J13,J14)

The above jumper settings correspond to selecting the on board microphone and the on board speaker, and also to use the on board audio amplifiers as shown in circuit diagram in Figure 6.26. Note that a microphone is connected to the MIC+ and MIC- inputs of the modem. Similarly, an audio amplifier is connected to pins SPKR+ and SPKR- of the modem and the amplifier drives a speaker. The audio amplifier in this project consists of an LM386 type audio chip with volume control.

We are now ready to establish connection to a mobile phone. The steps are given below:

- Select Aux audio channel (channel 1):

    AT+CHFA=1

- Select gain for the Aux microphone to 0dB (any other value up to 15 can be chosen):

    AT+CMIC=1,0

- Dial the mobile number required (in this example "07861688948"):

    ATD07861688948;

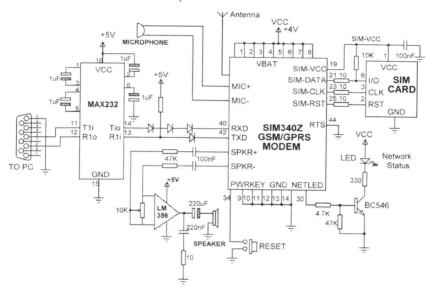

Figure 6.26 Circuit diagram of the project

The speaker should give an audible dialling tone and the mobile phone should ring. When the mobile phone answers, voice communication with the phone can be established.

To terminate the call, enter the AT command **ATH**. If the other side terminates the call then the modem responds with **NO CARRIER**.

Figure 6.27 shows an example where the above mobile phone number is dialled for voice communication.

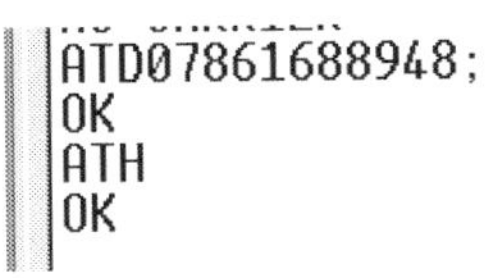

```
ATD07861688948;
OK
ATH
OK
```

Figure 6.27 Example dialling a mobile phone

6.9 GSM/GPRS – PC PROJECT 4 – ANSWERING A PHONE CALL

This section describes a project where a phone call is answered by the GSM/GPRS modem.

The SIM340 modem includes two microphones and two speaker connections. Smart GSM/GPSR board contains an on board speaker and an on board microphone that can be used for voice communication with a mobile phone. In addition, an external speaker and an external microphone can be connected to the Smart GSM/GPRS board. The Smart GSM/GPRS board contains audio amplifiers for the speaker and the microphone inputs. Although it is not essential to use audio amplifiers with the modem card, the signal levels can be increased by using audio amplifiers.

The Smart GSM/GPRS board jumper and switch settings are as in the previous project.

We are now ready to answer a phone call. The steps are given below:

- Select Aux audio channel (channel 1):

    AT+CHFA=1

- Select gain for the Aux microphone to 0dB (any other value up to 15 can be chosen):

    AT+CMIC=1,0

- You should see the message **RING** on the screen and the ringing tone should be heard on the speaker

- Answer the phone call by entering the following AT:

    ATA

You can now talk with the person at the other end of the phone. If you wish to terminate the call at any time enter the AT command **ATH**.

At any time during the call you can see the details of the caller by using the AT command AT+CLCC (see Figure 6.28).

```
AT+CLCC
+CLCC: 1,1,0,0,0,"07861688948",129,""

OK
```

Figure 6.28 Displaying the call details

6.10 GSM/GPRS – PC PROJECT 5 – SOME OTHER USEFUL COMMANDS

This section describes some other useful AT commands that can be used to find information about a call, about the SIM card, or about the mobile phone service provider.

6.10.1 Command AT+CLTS

This command displays the current date and time information. An example is shown in Figure 6.29 where the date is the 7th of March 2010 and the time is 10:00:15.

```
OK
AT+CLTS
+CLTS: "10/03/07,10:00:15+00"
OK
```

Figure 6.29 Command **AT+CLTS**

6.10.2 Command AT+CCVM

This command displays the voice mail number stored on the SIM card. An example is shown in Figure 6.30 where the number is "00447973100123"..

```
OK
AT+CCVM?
+CCVM: 1,"+447973100123",145,"Ans Phone"
OK
```

Figure 6.30 Command **AT+CCVM**

6.10.3 Command AT+CBAND

This command displays the current service provider operations frequency band. An example is given in Figure 6.31.

```
OK
AT+CBAND?
+CBAND: "EGSM_DCS_MODE"
OK
```

Figure 6.31 Command **AT+CBAND**

6.10.4 Command AT+CIMI

This command displays the International Mobile Subscriber Identity (IMI) of the SIM card. An example is shown in Figure 6.32 where the IMI number is "234332000453359".

```
AT+CIMI
234332000453359
OK
```

Figure 6.32 Command **AT+CIMI**

6.10.5 Command AT+CNUM

This command displays the phone number of the user (the SIM card issuing the command). An example is shown in Figure 6.33 where the phone number of the user issuing the command is "07971804172".

```
AT+CNUM
+CNUM: "","07971804172",129,7,4
OK
```

Figure 6.33 Command **AT+CNUM**

6.10.6 Command AT+CRSL

This command is used to set the ringer sound level (between 0 and 100). Figure 6.34 shows an example where the current setting is displayed as 100.

```
AT+CRSL?
+CRSL: 100

OK
```

Figure 6.34 Command **AT+CRSL**

6.10.7 Command AT+CLVL

This command is used to set the volume level of the speaker. Valid values are 0 to 100. Figure 6.35 shows an example where the current setting is displayed (70 in this example).

```
OK
AT+CLVL?
+CLVL: 70

OK
```

Figure 6.35 Command **AT+CLVL**

6.10.8 Command AT+CMUT

This command is used to turn the mute control ON (1) and OFF (0). Figure 6.36 shows how the mute control can be turned OFF.

```
OK
AT+CMUT=0
OK
```

Figure 6.36 Command **AT+CMUT**

CHAPTER 7

USING THE GSM/GPRS MODEM WITH MICROCONTROLLERS

7.1 Overview

In this chapter we shall be looking at how the GSM/GPRS modem card can be interfaced to a microcontroller and how we can send SMS messages.

The first project we shall be looking at shows how to send an SMS text message using a microcontroller and a modem card.

7.2 PROJECT 1 – SENDING AN SMS TEXT MESSAGE

In this project the SMS text message **"Hello from the microcontroller"** is sent to a mobile phone.

The block diagram of the project is shown in Figure 7.1. The microcontroller is connected to the GSM/GPRS modem card. The modem card powers and also drives the SIM card.

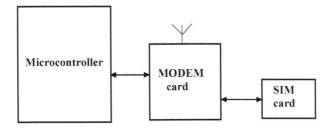

Figure 7.1 Block diagram of the project

The circuit diagram of the project is shown in Figure 7.2. A PIC16F887 type microcontroller is used in the project. The microcontroller interface is extremely simple. An 8MHz crystal is used as the clock source. The USART serial input pin RXD (RC7) and TXD (RC6) are connected to the TXD and RXD pins of the SIM340Z modem respectively using voltage level translation circuits. The RTS input of the modem is connected to ground. This pin should be forced Low when the modem is operated using only the TXD and RXD pins. The SIM card is connected to the modem via current limiting resistors where the modem provides the power supply voltage, I/O, CLK and RST inputs of the SIM card. The LED status output pin (NETLED) of the modem is connected to an LED through a transistor switch circuit. An RF antenna supplied with the modem is connected to the antenna socket of the modem. The microcontroller can be reset via its MCLR input using an external push-button switch. Similarly, the modem can be reset via its PWRKEY input using an external switch. The PWRKEY input should be held Low for at least four seconds to re-start the modem. It is important to note that the modem operating voltage is between 3.5V and 4.5V, with a recommended typical value of 4V. The power supply should be capable of providing up to 2A current which may be required in transient operations.

The system was constructed using an *EasyPIC 5* microcontroller development board together with a *Smart GSM/GPRS* development board. The SIM340Z modem card is mounted on the GSM/GPRS development board (to the bottom center part) with its antenna attached at the side of the board. PORT C of the EasyPIC 5 development board was connected to modem terminals of the GSM/GPRS board via a 10-way ribbon cable, as in Figure 7.3. Jumper J1 on the GSM/GPRS board was set to select communication with an MPU. The EasyPIC 5 board was powered from the USB port of a PC (laptop), while the GSM/GPRS board was powered from an external 12V mains adaptor.

The software was developed using the mikroC compiler. mikroC language supports a large variety of interface devices and protocols, and provides built-in library of functions for devices such as SD card, CompactFlash card, I^2C bus, RS232 and RS485, LCD, USB, CAN bus and so on. Figure 7.4 shows operation of the software using a PDL (Program Description Language) type description. At the beginning of the program various constant character strings such as the AT commands and mobile phone number used in the program are declared. The main program forces the modem RTS pin to 0, configures the analog input channel AN0 and enables the USART interrupts. The USART is initialised with a Baud Rate of 19200 and a loop is formed to force the modem Baud Rate to this value. The program then sets the modem into Text mode with GSM character code, 24 hrs message validity period, and the message class of 0. Then the program sends the SMS message "**Hello from the microcontroller**" and terminates.

USART interrupts are used to receive the modem responses. A response can be "OK" or "RDY" and is terminated with a carriage-return and line-feed character pair. After sending an AT command, the program waits until a successful response is received from the modem.

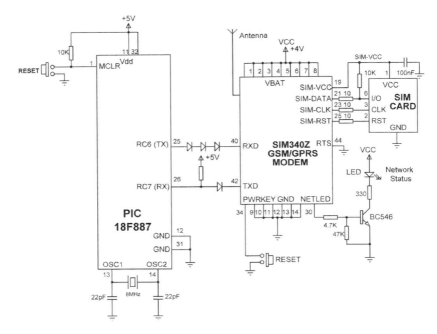

Figure 7.2 Circuit diagram of the project

Figure 7.3 Connecting the EasyPIC 5 board to the Smart GSM/GPRS board

MAIN PROGRAM

BEGIN
    Configure digital I/O ports
    Configure USART interrupts
    Set USART to 19200
    Set Modem to Auto-baud
    Disable Echo
    Set modem to text mode
    Set character mode to GSM
    Send SMS "Hello from the microcontroller"
END

SEND SMS

BEGIN
    Set modem CSMP parameters
    Set mobile number
    Send SMS
    Send Ctrl-z character
END

Figure 7.4 PDL of the program

The full program listing of the project is given in Figure 7.5.

```
/**********************************************************************
                        SENDING SMS MESSAGE
                        ===================
```

In this project a text message is sent to a mobile phone. A PIC16F887 microcontroller, operated with a 8MHz crystal is used in the project. A SIM340Z type GSM/GPRS modem card is connected to the serial ports of the microcontroller. The microcontroller sends the following text message as an SMS:

    Hello from the microcontroller

Author: Dogan Ibrahim
Date: March 2010
File: SIM.C

```
**********************************************************************/
```

```c
#define OK   0
#define RDY  1

char AT[] = "AT";
char NoEcho[] = "ATE0";
char Mode_Text[] = "AT+CMGF=1";
char Ch_Mode[] = "AT+CSCS=\"GSM\"";
char Param[] = "AT+CSMP=17,167,0,241";
char Mobile_No[] = "AT+CMGS=\"07528885664\"";
char Terminator = 0x1A;
char msg[] = "Hello from the microcontroller";

char State = 0;
char response_rcvd = 0;
short responseID = -1, response = -1;


short Modem_Response()
{
    if(response_rcvd)
    {
       response_rcvd = 0;
       return responseID;
    }
    else return -1;
}



//
// This function waits for a modem response
//
void Wait_Modem_Response(char resp)
{
    while(Modem_Response() != resp);
}



//
// This function sends a character string to the modem
//
void Send_To_Modem(char *s)
{
    while(*s) Usart_Write(*s++);           // Send Cmd
    Usart_Write(0x0D);                      // Send CR
}


//
```

```c
// Interrupt Service Routine
//
void interrupt()
{    char Dat;

    if(PIR1.RCIF == 1)
    {
      Dat = Usart_Read();
      switch(State)
      {
       case 0:
           {
             response = -1;
             if(Dat == 'O')State = 1;
             if(Dat == '>')
             {
              response_rcvd = 1;
              ResponseID = RDY;
              State = 0;
             }
             break;
           }

        case 1:
            {
              if(Dat == 'K')
              {
               response = OK;
               State = 2;
              }
              break;
            }

        case 2:
            {
              if(Dat == 0x0D)
                State = 3;
              else State = 0;
              break;
            }

        case 3:
            {
              if(Dat == 0x0A)
              {
                response_rcvd = 1;
                ResponseID = response;
              }
              State = 0;
              break;
```

```
            }
        default:
            {
                State = 0;
                break;
            }
        }
    }
}


//
// This function sends SMS message using the GSM/GPRS modem
//
void Send_SMS(void)
{
//
// Send Parameters
//
    Send_To_Modem(Param);
    Wait_Modem_Response(OK);
//
// Mobile phone number
//
    Send_To_Modem(Mobile_No);
    Wait_Modem_Response(RDY);
    Send_To_Modem(Msg);
    USART_Write(Terminator);
    Wait_Modem_Response(OK);
}


//
// Start of MAIN program
//
void main()
{
    TRISC = 0x80;
//
// Enable USART interrupt
//
    PIE1.RCIE = 1;
    INTCON.PEIE = 1;
    INTCON.GIE = 1;

    Usart_Init(19200);                  // Initialize USART
    Delay_Ms(5000);                     // Wait for GSM module
```

```
    while(1)                              // Auto-baud detect
    {
      Send_To_Modem(AT);
      Delay_Ms(100);
      if(Modem_Response() == OK)break;
    }
//
// Disable Echo
//
   Send_To_Modem(NoEcho);
   Wait_Modem_Response(OK);
//
//Set message mode to TEXT
//
   Send_To_Modem(Mode_Text);
   Wait_Modem_Response(OK);
//
// Set character mode to GSM
//
   Send_To_Modem(Ch_Mode);
   Wait_Modem_Response(OK);
//
// Send SMS
//
    Send_SMS();
//
// End of program. Wait here forever
//
     for(;;);
}
```

Figure 7.5 Program listing of the project

7.3 PROJECT 2 – SENDING THE TEMPERATURE AS SMS TEXT MESSAGES

In this project the ambient temperature is read every 10 minutes and is then sent to a specified mobile phone as SMS text messages.

7.3.1 The Block Diagram

The block diagram of the designed system is shown in Figure 7.6. A semiconductor analog sensor is used as the temperature sensor. The microcontroller converts the sensor output voltage from analog to digital. The GSM/GPRS modem is under the control of the microcontroller. Standard AT commands are used to communicate with the modem and to send the temperature as an SMS message to a mobile phone. A SIM card provides the subscriber details to the system. More information about various parts of the system are given in the following sections.

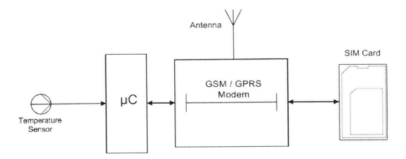

Figure 7.6 Block diagram of the project

7.3.2 The Temperature Sensor

A 3-pin LM35DZ type analog semiconductor temperature sensor is used in the design. This sensor can be used to measure temperatures in the range 0°C to 70°C. Figure 7.7 shows a picture of the sensor. The output voltage of the sensor is given by Vo = 10mV/°C. Thus, for example, at 10°C the sensor output voltage is 100mV. Similarly, at 30°C the output voltage is 300mV.

Figure 7.7. LM35DZ temperature sensor

7.3.3 The Circuit Diagram

The circuit diagram of the designed system is shown in Figure 7.8. The microcontroller interface is extremely simple. An 8MHz crystal is used as the clock source. The temperature sensor is connected to analog input AN0 of the microcontroller. USART serial input pin RXD (RC7) and TXD (RC6) are connected to the TXD and RXD pins of the SIM340Z modem respectively via voltage level translator circuits (3.3V to 5.0 and vice-versa). The RTS input of the modem is connected to ground as it is not used. The SIM card is connected to the modem via current limiting resistors where the modem provides the power supply voltage, I/O, CLK and RST inputs of the SIM card. The LED status output pin (NETLED) of the modem is connected to an LED through a transistor switch circuit. An RF antenna supplied with the modem is connected to the antenna socket of the modem. The microcontroller can be reset via its MCLR input using an external push-button switch. Similarly, the modem can be reset via its PWRKEY input using an external switch. The PWRKEY input should be held Low for at least four seconds to re-start the modem. It is important to note that the modem operating voltage is between 3.5V and 4.5V, with a recommended typical value of 4V. The power supply should be capable of providing up to 2A current which may be required in transient operations.

The following switch settings should be set when using the Smart GSM/GPRS board with a microcontroller:

    Set SELECT UART JUMPER to MCU
    Set switches ENABLE UART INTERFACE TO PC to OFF

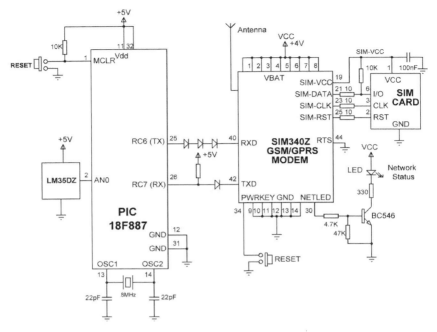

Figure 7.8 Circuit diagram of the system

7.3.4 The Software

The software was developed using the mikroC compiler. This is a popular microcontroller C language compiler developed by mikroElektronika (www.mikroe.com). mikroC language supports a large variety of interface devices and protocols, and provides built-in library of functions for devices such as SD card, CompactFlash card, I^2C bus, RS232 and RS485, LCD, USB, CAN bus and so on. Figure 7.9 shows operation of the software using a PDL (Program Description Language) type description. At the beginning of the program the digital and analog I/O ports are configured, various constant character strings such as the AT commands and mobile phone number used in the program are declared. The main program configures the analog input channel AN0 and enables the USART interrupts. The USART is initialised with a Baud Rate of 19200 and a loop is formed to force the modem Baud Rate to this value. The program then sets the modem into Text mode with GSM character code, 24 hrs message validity period, and the message class of 0. Then the program enters an endless loop where the temperature is read every hour and sent to the specified mobile phone as an SMS text message. The temperature is read from analog channel AN0, converted into millivolts, and then divided by 10 to find the actual physical temperature in °C.

USART interrupts are used to receive the modem responses. A response can be "OK" or "RDY" and is terminated with a carriage-return and line-feed character pair. After sending an AT command, the program waits until a successful response is received from the modem.

The software developed in this project can be improved by the following modifications:

- Several temperature readings can be collected and then sent at the same time using one SMS message.
- The SMS messages can be sent in response to a request from the recipient. Thus data can be collected and received whenever required.
- A flash memory (e.g. an SD card) can be added to store the collected data.
- GPRS mode or FTP can be used to send large amount of data.
- An RTC chip can be added to the system to timestamp the collected data.

MAIN PROGRAM:

BEGIN
    Configure digital I/O ports
    Configure analog channel
    Configure USART interrupts
    Set USART to 19200 Baud
    Force RTS Low
    Set Modem to Auto-baud
    Disable Echo
    Set Modem to text mode
    Set character mode to GSM
    DO FOREVER
        Read temperature()
        Send SMS()
        Wait one hour
    ENDDO
END

Send SMS:

BEGIN
    Set Modem CSMP parameters
    Set mobile phone number
    Send SMS (temperature)
    Send Ctrl-Z character
END

Read Temperature:

BEGIN
    Read analog channel AN0
    Convert to mV
    Convert to °C
    Store as message
END

Figure 7.9 PDL of the project

The full program listing of the project is given in Figure 7.10.

```
/***********************************************************************
                GSM/GPRS TEMPERATURE DATA COLLECTION
                ====================================
```

This is the temperature data collection program with SMS output. Temperature data is read every 10 minutes from a LM35DZ type sensor and then sent to a mobile phone every 10 minutes as SMS messages.

In this project a PIC16F887 type microcontroller is used. The microcontroller is operated from a 8MHz crystal.

The temperature sensor is connected to analog port AN0 of the microcontroller. The serial output (TXD) and serial input (RXD) of the microcontroller are connected to the SIM340 type modem via voltage translator circuits. The modem drives a SIM card directly.

Author: Dogan Ibrahim
Date: March 2010
File: GSM.C

```
***********************************************************************/

#define OK   0
#define RDY  1

char AT[] = "AT";
char NoEcho[] = "ATE0";
char Mode_Text[] = "AT+CMGF=1";
char Ch_Mode[] = "AT+CSCS=\"GSM\"";
char Param[] = "AT+CSMP=17,167,0,241";
char Mobile_No[] = "AT+CMGS=\"07528885664\"";
char Msg[] = "Temperature =           ";
char Terminator = 0x1A;
const int lsb = 5000/1024;
int mV,temperature,Vin;
```

```
char State = 0;
char response_rcvd = 0;
short responseID = -1, response = -1;

//
// This function processes the modem responses
//
short Modem_Response()
{
    if(response_rcvd)
    {
     response_rcvd = 0;
     return responseID;
    }
    else return -1;
}



void Wait_Modem_Response(char resp)
{
    while(Modem_Response() != resp);
}



//
// This function sends a character string to the Modem
//
void Send_To_Modem(char *s)
{
    while(*s) Usart_Write(*s++);          / Send Cmd
    Usart_Write(0x0D);                    // Send CR
}



//
// Interrupt Service Routine
//
void interrupt()
{   char Dat;

    if(PIR1.RCIF == 1)
    {
      Dat = Usart_Read();
      switch(State)
      {
       case 0:
          {
```

```
            response = -1;
            if(Dat == 'O')State = 1;
            if(Dat == '>')
            {
                response_rcvd = 1;
                ResponseID = RDY;
                State = 0;
            }
            break;
            }

        case 1:
            {
            if(Dat == 'K')
            {
              response = OK;
              State = 2;
            }
            break;
            }

        case 2:
            {
             if(Dat == 0x0D)
               State = 3;
             else State = 0;
             break;
            }

        case 3:
            {
             if(Dat == 0x0A)
             {
                 response_rcvd = 1;
                 ResponseID = response;
             }
             State = 0;
             break;
            }

        default:
            {
             State = 0;
             break;
            }
    }
  }
}
```

```c
//
// This function reads the temperature
//
void Read_Temperature(void)
{
    Vin = Adc_Read(0);                  // Read analog data
    mV = Vin * lsb;                     // Convert to mV
    temperature = mV / 10;              // Convert to temperature
    IntToStr(temperature, Msg+14);
}


//
// This function sends send SMS messages to the Modem
//
void Send_SMS(void)
{
//
// Send Parameters
//
    Send_To_Modem(Param);
    Wait_Modem_Response(OK);
//
// Mobile phone number
//
    Send_To_Modem(Mobile_No);
    Wait_Modem_Response(RDY);
    Send_To_Modem(Msg);
    USART_Write(Terminator);
    Wait_Modem_Response(OK);
}


//
// This function creates 10 minutes delay
//
void Wait_Ten_Minutes()
{
    int i;
    for(i=0; i < 600; i++)Delay_Ms(1000);
}


//
// START OF MAIN PROGRAM
//
void main()
{
    TRISC = 0x80;
```

```
    TRISA = 1;
    ANSEL = 1;
    ADCON1 = 0x80;                          // Use AN0 ith Vref=+5V
//
// Enable USART interrupt
//
    PIE1.RCIE = 1;
    INTCON.PEIE = 1;
    INTCON.GIE = 1;

    Usart_Init(19200);                      // Initialize USART
    Delay_Ms(5000);                         // Wait for GSM module

    while(1)                                // Auto-baud detect
    {
      Send_To_Modem(AT);
      Delay_Ms(100);
      if(Modem_Response() == OK)break;
    }
//
// Disable Echo
//
    Send_To_Modem(NoEcho);
    Wait_Modem_Response(OK);
//
//Set message mode to TEXT
//
    Send_To_Modem(Mode_Text);
    Wait_Modem_Response(OK);
//
// Set character mode to GSM
//
    Send_To_Modem(Ch_Mode);
    Wait_Modem_Response(OK);
//
// Read the temperature every 10 minutes and send as SMS
//
    for(;;)
    {
        Read_Temperature();
        Send_SMS();
        Wait_Ten_Minutes();
    }
}
```

Figure 7.10 Program listing of the project

Figure 7.11 shows the message sent to a mobile phone.

Figure 7.11 Message sent to a mobile phone

7.4 PROJECT 3 – SENDING SMS USING A MICROCONTROLLER AND A PC

In this project a PC is connected to the microcontroller using the serial port. The user is prompted to enter a mobile phone number and an SMS message. The message is then sent to the specified mobile phone.

The block diagram of the project is shown in Figure 7.12.

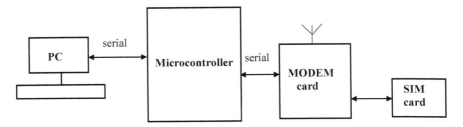

Figure 7.12 Block diagram of the project

The circuit diagram of the project is shown in Figure 7.13. Two serial ports are required for the project: one for the interface with the PC, and the other one for the interface with the Modem card.

The PC serial port is connected to microcontroller port pins RB1 and RB5 through a MAX232 type voltage converter chip. Pin RB1 is used as serial output (TX), and pin RB5 is used as serial input (RX). mikroC software serial library routines are used to control these pins.

The microcontroller (PIC16F887) is connected to the modem via the USART module of the microcontroller and through voltage translator circuits. Pin RC6 is the serial output (TX), and pin RC7 is the serial input (RX) of the microcontroller.

The modem card drives the SIM card and provides supply voltage to the card.

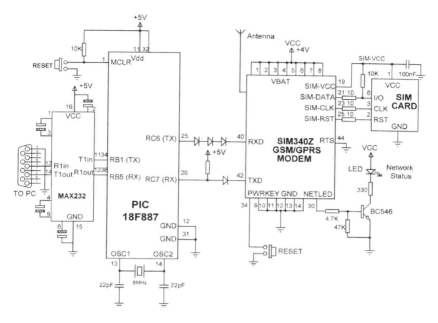

Figure 7.13 Circuit diagram of the project

If using the EasyPIC 5 development board and the Smart GSM/GPRS board, PORT C of the EasyPIC 5 board (10 way connector at the edge) should be connected to the PIC UART interface of the Smart GSM/GPRS board. In addition, the following are required on each board:

Smart GSM/GPRS board:

    ENABLE UART INTERFACE TO PC switches should be OFF
    SELECT UART INTERFACE should be set to MCU

EasyPIC 5 board:

    SW7 switch 6 should be ON
    SW8 switch 4 should be ON
    PC serial cable should be connected to the serial connector

The PDL of the project is shown if Figure 7.14. The main program configures the PORTC I/O pin directions. ANSELH register is then cleared to 0 so that PORTB pins are digital I/O. USART interrupts are then enabled by setting:

PIE1.RCIE = 1
INTCON.PIEI = 1
INTCON.GIE = 1

The USART is initialized to 19200 baud. The program then waits for 5 seconds for the modem to be ready. The modem baud rate is set to 19200 by using an auto baud-detect software. Here, the "AT" command is sent to the modem continuously until the modem responds with "OK". Then the echo mode is disabled, modem is set into text SMS mode and the character set is set to "GSM". The program then configures and initializes the soft USART library where port pins RB1 and RB5 are configured as serial input and output. The rest of the program is executed in an endless loop formed using a **for** statement. Inside this loop the following message is displayed on the PC screen:

GSM/GPRS SYSTEM – SEND SMS
============================

Enter Phone No:

The user enter the mobile phone number where the message is to be sent to and presses ENTER. The program then prompts:

Message to Send:

The user enters the message to be sent and presses the ENTER key. The program displays:

Sending the message...

The message is sent to the specified mobile phone and then the program displays:

Message sent...

The above process is repeated until the user has no more messages to be sent.

GSM/GPRS SYSTEM – SEND SMS
============================

Enter Phone No:

BEGIN
    Configure digital I/O ports
    Configure USART interrupts
    Set USART to 19200
    Set Modem to Auto-baud
    Disable Echo
    Set modem to text mode
    Set character mode to GSM
    Initialize Soft Serial Port Library (Soft USART)
    DO FOREVER
        Display "GSM/GPRS SYSTEM – SEND SMS"
        Display "Enter Phone No:"
        Read Phone number
        Display "Message to Send:"
        Read the message
        Display "Sending the message…"
        Display "Message sent…"
    ENDDO
END

Figure 7.14 PDL of the project

The program listing of the project is given in Figure 7.15.

```
/***********************************************************************
                        SENDING SMS MESSAGE
                        ====================
```

In this project a PC is connected to the project. The user is prompted to enter a mobile phone number and a message. The message is sent to the given number.

Author: Dogan Ibrahim
Date: March 2010
File: SIMPC.C

```
***********************************************************************/

#define OK  0
#define RDY 1

char AT[] = "AT";
char NoEcho[] = "ATE0";
char Mode_Text[] = "AT+CMGF=1";
char Ch_Mode[] = "AT+CSCS=\"GSM\"";
char Param[] = "AT+CSMP=17,167,0,241";
char Mobile_No[] = "AT+CMGS=\"         \"";
```

```c
char Terminator = 0x1A;
char sms_msg[50];
char State = 0;
char response_rcvd = 0;
short responseID = -1, response = -1;


//
// Send NULL terminated TEXT message to Soft UART
//
void Send_Text(const char mesg[])
{
  char k;
  k = 0;
   while(mesg[k] != 0x0)
   {
     Soft_Uart_Write(mesg[k]);
     k++;
   }
}


//
// Send Newline (carriage-return and line-feed) to Soft UART
//
void Send_Newline(void)
{
  Soft_Uart_Write(0x0A);
  Soft_Uart_Write(0x0D);
}


//
// Read from Soft UART until a carriage-return is detected
//
void Read_Soft_Uart(unsigned char usart[])
{
 unsigned char k,flag;
 unsigned char dat,*rec;

 k = 0;
 flag = 1;

 while(flag == 1)
  {
    do
    {
      dat = Soft_Uart_Read(rec);
    }while(*rec);
    usart[k] = dat;
```

```c
        Soft_Uart_Write(dat);
        if(dat == 0x0D)flag = 0;
        k++;
    }
}


short Modem_Response()
{
    if(response_rcvd)
    {
        response_rcvd = 0;
        return responseID;
    }
    else return -1;
}


//
// This function waits for a modem response
//
void Wait_Modem_Response(char resp)
{
    while(Modem_Response() != resp);
}


//
// This function sends a character string to the modem
//
void Send_To_Modem(char *s)
{
    while(*s) Usart_Write(*s++);            // Send Cmd
    Usart_Write(0x0D);                      // Send CR
}


//
// Interrupt Service Routine
//
void interrupt()
{   char Dat;

    if(PIR1.RCIF == 1)
```

```c
{
    Dat = Usart_Read();
    switch(State)
    {
      case 0:
          {
            response = -1;
            if(Dat == 'O')State = 1;
            if(Dat == '>')
            {
             response_rcvd = 1;
             ResponseID = RDY;
             State = 0;
            }
            break;
          }

      case 1:
          {
           if(Dat == 'K')State = 2;
           break;
          }

      case 2:
          {
           if(Dat == 0x0D)
             State = 3;
           else State = 0;
           break;
          }

      case 3:
          {
           if(Dat == 0x0A)
           {
            response_rcvd = 1;
            ResponseID = OK;
           }
           State = 0;
           break;
          }

      default:
          {
           State = 0;
           break;
          }
    }
  }
}
```

```
//
// This function sends SMS message using the GSM/GPRS modem
//
void Send_SMS(void)
{
//
// Send Parameters
//
    Send_To_Modem(Param);
    Wait_Modem_Response(OK);
//
// Mobile phone number
//
    Send_To_Modem(Mobile_No);
    Wait_Modem_Response(RDY);
    Send_To_Modem(sms_msg);
    USART_Write(Terminator);
    Wait_Modem_Response(OK);
}


//
// Start of MAIN program
//
void main()
{

    TRISC = 0x80;
    ANSELH = 0;
    TRISB = 0x20;

//
// Enable USART interrupt
//
    PIE1.RCIE = 1;
    INTCON.PEIE = 1;
    INTCON.GIE = 1;

    Usart_Init(19200);                   // Initialize USART
    Delay_Ms(5000);                      // Wait for GSM module

    while(1)                             // Auto-baud detect
    {
      Send_To_Modem(AT);
```

```
        Delay_Ms(100);
        if(Modem_Response() == OK)break;
    }
//
// Disable Echo
//
    Send_To_Modem(NoEcho);
    Wait_Modem_Response(OK);
//
//Set message mode to TEXT
//
    Send_To_Modem(Mode_Text);
    Wait_Modem_Response(OK);
//
// Set character mode to GSM
//

    Send_To_Modem(Ch_Mode);
    Wait_Modem_Response(OK);

//
// Initialize soft UART. RX=RB5 and TX=RB1
//
// Display MENU and get a choice
//
    Soft_Uart_Init(PORTB, 5, 1, 2400, 0);

   for(;;)
   {
     Send_Newline();
     Send_Newline();
     Send_Text("GSM/GPRS SYSTEM - SEND SMS");
     Send_Newline();
     Send_Text("=========================");
     Send_Newline();
     Send_Newline();
     Send_Text("Enter Phone No: ");
     Read_Soft_Uart(&Mobile_No[9]);
     Send_Newline();
     Send_Text("Message to Send: ");
     Read_Soft_Uart(sms_msg);
     Send_Newline();
     Send_Text("Sending the message...");
     Send_Sms();
     Send_Newline();
     Send_Text("Message sent...");
   }

}
```

Figure 7.15 Program listing of the project

Figure 7.16 shows a typical dialogue where a user defined message is sent to a given mobile phone.

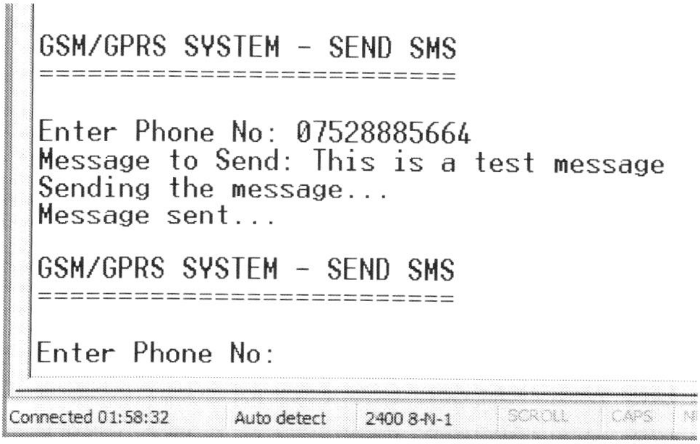
GSM/GPRS SYSTEM - SEND SMS
==========================
Enter Phone No: 07528885664
Message to Send: This is a test message
Sending the message...
Message sent...
GSM/GPRS SYSTEM - SEND SMS
==========================
Enter Phone No:
Connected 01:58:32
Auto detect
2400 8-N-1
SCROLL
CAPS

Figure 7.16 Sending an SMS to a mobile phone

CHAPTER 8

THE REAL TIME CLOCK CHIP

8.1 Overview

The real-time clock (RTC) is very important in real-time data logging applications as it enables the data to be time-stamped with the current date and time information.

This Chapter describes the basic operating principles and interfacing techniques of one of the popular RTC chips in microcontroller based projects. The RTC chip used in this Chapter is the PCF8583 clock calendar chip[8] with built-in 256 x 8 static RAM.

8.2 The PCF8583 RTC Chip

PCF8583 is an I^2C protocol based 8-pin small integrated circuit which is used in microcontroller based systems to provide the real time date and time information. In addition to the normal clock functions, the chip also incorporates a small built-in RAM memory.

Figure 8.1 shows pin configuration of the chip. Pins 1 and 2 are the oscillator pins and a crystal with a frequency of 32768Hz should be connected between these pins. Pin 3 is the chip address pin and is normally connected to ground if there is only one similar device in the system. Pin 4 is the ground pin. Pins 5 and 6 are the I^2C data and clock functions respectively. Pin 7 is a pulse output with a rate of 1 second and can be used in microcontroller circuits as an accurate one second pulses. Pin 8 is the power supply pin which is usually connected to a +5V supply during normal operation.

PCF8583 has the following features:

- 2.5V to 6.0V operating voltage
- I^2C bus compatible
- 12 or 24 hour time format
- programmable alarm indication
- on chip 256 x 8 RAM memory
- crystal input for accurate timing

Figure 8.1 Pin configuration of PCF8583

The first 8 locations of the RAM (address 0 to 7) are reserved for clock functions. Address 0 contains a control and status register and addresses 1 to 7 store the clock data. RAM addresses 8 to 255 are available as general purpose registers or may be programmed as alarm registers.

8.2.1 The Control and Status Register

The control and status register (CSR) controls all functions of the chip. In this book we are only concerned with the clock functions of the chip. Bits 4 and 5 of the CSR control the clock mode as follows:

```
00    32768Hz clock
01    50Hz clock
10    event counter mode
11    test mode
```

In normal applications a 32768Hz crystal is used and the 32768Hz clock mode is selected (default mode). Bit 7 of the CSR is used to stop and start the clock. When 1, clock is stopped, and when 0 the clock counts.

8.2.2 The Counter Registers

The counter registers store the real time date and clock information. Figure 8.2 shows the layout of these registers. The date and time are stored in BCD format.

Table 8.1 shows the valid values of various date and time registers. As it can be seen from this table the hours field can either be in 24-hour format (00 – 23 hrs), or in 12 hour format (01-12 with a.m. and p.m fields).

	1/100 second	
1/10s		1/100s
	Seconds	
10s		1s
	Minutes	
10min		1min
	Hours	
10hr		1hr
	Year/date	
10day		1day
	Weekday/month	
10month		1month
	Timer	
10day		1day

Figure 8.2 The counter registers

Table 8.1 Valid values of date and clock counter registers

1/100 second	00 - 99
Second	00 – 59
Minute	00 – 59
Hour (24hrs format)	00 – 23
Hour (12 hrs format)	01AM – 12PM
Day	01 – 31
Month	01 – 12
Year	0 – 3
Weekday	0 – 6
Timer	00 – 99

8.2.3 Format of the Hours Register

The hours register (at address 4 of the RAM) has special format and is shown in Figure 8.3. Bits 0 – 3 are the units of the Hours field, bits 4 and 5 are the tens of the Hours field. Bit 6 indicates the AM or PM of the hour. Bit 7 should be set to 0 for 24 Hrs operation mode, and to 1 for 12 Hrs operation mode.

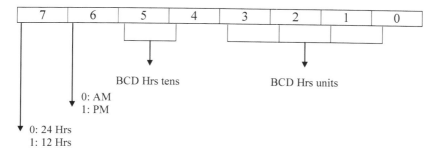

Figure 8.3 Format of the Hours register

8.2.4 Format of the Year/Date Register

The year/date register (at address 5 of the RAM) has special format and is shown in Figure 8.4. Bits 0 – 3 are units of the Days field, bits 4 and 5 are tens of the hours field. Bits 6 and 7 specify the year field. This field can take values from 0 to 3 and is updated on the new year. There are several ways that these bits can be used to represent the year. One way is to store the actual year in one of the RAM locations of the chip (or in the EEPROM memory of the microcontroller) and then use the Year bits to update the stored year value. For example, if the year value "08" is stored in a RAM location with a "0" stored in the year field, then when the year field becomes a "1" the actual year in the RAM can be updated to "09". Alternatively, the year value stored in the RAM can be adjusted every 4 years using the four values of the Year bits.

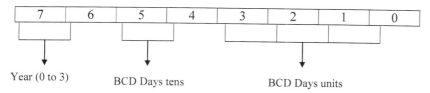

Figure 8.4 Format of the Year/Date register

8.2.5 Format of the Weekdays/Month Register

The weekdays/month register (at address 6 of the RAM) has special format and is shown in Figure 8.5. Bits 0 – 3 are the units of the Month field, bit 4 is the tens of the month field. Bits 5, 6 and 7 are the weekdays.

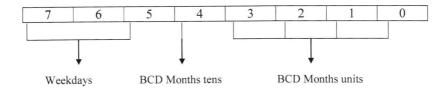

Figure 8.5 Format of the Weekdays/Month register

8.3 Using the PCF8583 RTC Chip in Microcontroller Projects

This section describes how the PCF8583 RTC chip can be used in microcontroller based projects.

Figure 8.6 shows how the chip can be interfaced to a PIC microcontroller. Pins SCL and SDA of the RTC chip use the I^2C protocol and they must be pulled up with suitable resistors (4.7K to 10K). Notice the connection between the RTC chip and the microcontroller: pin SDA of the RTC chip is connected to pin RC3 of the microcontroller, and pin SCL of the RTC chip is connected to pin RC4 of the microcontroller. These are the recommended connections when the **mikroC** language I^2C library functions are used to communicate with devices over the I^2C bus.

A 32768Hz crystal is used to provide accurate clock pulses to the RTC chip and the timing can be adjusted accurately using a variable capacitor. A battery back-up is provided to supply power to the chip so that the date and time information are not lost when power is removed from the chip. The battery circuit is protected by two diodes and is connected to the main power supply (+5V) of the microcontroller.

8.3.1 Writing to the PCF8583 RTC Chip

The date and time stored on the RTC chip should be set to the current correct date and time when the chip is used the first time, or when it is required to change the date and/or the time stored on the chip.

The following steps should be followed to write new date and time information to the RTC chip.

- Start I²C bus
- Send address for writing (0xA0)
- Write seconds
- Write minutes
- Write hours
- Write day and year
- Write month
- Stop I²C bus
- Restart I²C bus
- Send address for writing (0xA0)
- Start counter
- Stop I²C bus

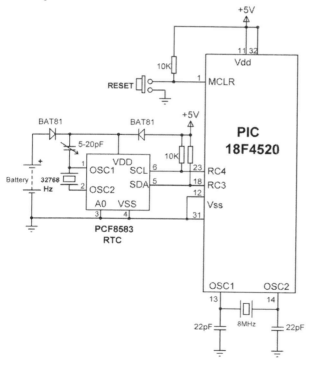

Figure 8.6 Interfacing the RTC chip to a PIC microcontroller

8.3.2 Reading From the PCF8583 RTC Chip

The following steps should be followed to read the date and time from the RTC chip:

- Start I²C bus
- Send address for writing (0xA0)
- Send address of the seconds register (0x02)
- Restart the I²C bus
- Send address for reading (0xA1)
- Read seconds
- Read minutes
- Write hours
- Read day and year
- Calculate the actual years
- Read month
- Stop I²C bus

More details on how to program the RTC chip for reading and writing are given in the project program section of the book.

8.4 Example Reading From the RTC Chip

An example is given in this section to show how the date and clock data can be read from the RTC chip. The circuit diagram for this example is as in Figure 8.6 where the RTC chip is connected to PORT C pins RC3 and RC4 of a PIC18F4520 type microcontroller, operated at 8MHz (any other type of microcontroller can also be used). The code given below (Figure 8.7) reads the date and time and stores in arrays RTCDate and RTCTime respectively. The code uses the soft I²C library routines of the mikroC language. The date is stored in the following format in array RTCDate:

dd-mm-yy

Similarly, the time is stored in the following format in array RTCTime:

hh:mm:ss

RTC date and clock data are stored in BCD format inside the RTC chip. Macro routines MSB and LSB read the MSB and LSB fields of the date and clock data and convert into ASCII format. In this code the year field is stored in EEPROM address 0 of the microcontroller:

```
#define MSB(x) ((x >> 4) + '0')
#define LSB(x) ((x & 0x0F) + '0')

unsigned char rtc,year,seconds;

Soft_I2C_Config(&PORTC, 3, 4);              // RC3=SDA, RC4=SCL

Soft_I2C_Start();                            // Start I2C Bus
Soft_I2C_Write(0xA0);                        // Address PCF8583
Soft_I2C_Write(2);
Soft_I2c_Start();
Soft_I2C_Write(0xA1);                        // Address for reading
rtc = Soft_I2C_Read(1);                      // Read seconds
RTCTime[6] = MSB(rtc);
RTCTime[7] = LSB(rtc);
seconds=10*(RTCTime[6]-'0')+RTCTime[7]-'0';
new_time = seconds;
rtc = Soft_I2C_Read(1);                      // Read minutes
RTCTime[3] = MSB(rtc);
RTCTime[4] = LSB(rtc);
rtc = Soft_I2C_Read(1);                      // Read hours
RTCTime[0] = MSB(rtc);
RTCTime[1] = LSB(rtc);
rtc = Soft_I2C_Read(1);                      // Read year/day
RTCDate[0] = MSB((rtc & 0x30));
RTCDate[1] = LSB(rtc);
year = (rtc & 0xC0) >> 6;                    // Year starts from 2008
rtc = Soft_I2C_Read(0);                      // Read weekday/month
RTCDate[3] = MSB((rtc & 0x10));
RTCDate[4] = LSB(rtc);
Soft_I2C_Stop();                             // Stop I2C bus

//
// Year adjustment. The year is stored in the first location of the EEPROM memory
// of the PIC microcontroller
//

rtc = EEPROM_Read(0);                        // EEPROM 0 stores years
if(year != 0)
{
      rtc++;
      EEPROM_WRITE(0,rtc);
}
RTCDate[6] = (rtc / 10) + '0';
RTCDate[7] = (rtc % 10) + '0';
```

Figure 8.7 Reading from the RTC chip

8.5 Example Writing to the RTC Chip

An example is given in this section to show how new date and clock data can be written to the RTC chip. The circuit diagram is as in Figure 8.6. The ASCII date and time data are converted into BCD format (see Figure 8.8) using Macro routine CLKD and then stored in the appropriate fields of the RTC chip. It is assumed that the date and time were stored in arrays RTCDate and RTCTime respectively. It is assumed that address 0 of the EEPROM stores the year field:

```
#define CLKD(x,y) (((x - '0') << 4) + y - '0')

Soft_I2C_Config(&PORTC, 3, 4);              // RC3=SDA, RC4=SCL

Soft_I2C_Start();                            // Start Soft I2C
Soft_I2C_Write(0xA0);
Soft_I2C_Write(0);
Soft_I2C_Write(0x80);
Soft_I2C_Write(0);
Soft_I2C_Write(CLKD(RTCTime[6],RTCTime[7])); // Seconds
Soft_I2C_Write(CLKD(RTCTime[3],RTCTime[4])); // Minutes
Soft_I2C_Write(CLKD(RTCTime[0],RTCTime[1])); // Hours

year = 10*(RTCDate[6] - '0') + RTCDate[7] - '0';
EEPROM_Write(0,year);
yr = (year % 4) << 2;
RTCDate[0] = RTCDate[0] + yr;

Soft_I2C_Write(CLKD(RTCDate[0],RTCDate[1])); // Day
Soft_I2C_Write(CLKD(RTCDate[3],RTCDate[4])); // Month
Soft_I2C_Stop();
Soft_I2C_Start();

Soft_I2C_Write(0xA0);
Soft_I2C_Write(0);
Soft_I2C_Write(0);                           // Enable counting
Soft_I2C_Stop();
```

Figure 8.8 Writing to the RTC chip

REFERENCES

1. Thermo Recorder TR-5 Series Data Sheet
   Web site: www.tdlogger.com/dloggers/Lit/datashts/TR-5.pdf

2. DrDaq Data Logger
   Web site: http://www.picotech.com

3. Microchip
   Web site: http://www.microchip.com

4. mikrolektronika
   Web site: http://www.mikroe.com

5. Hi-tech Software
   Web site: http://www.htsoft.com

6. CCS Compiler
   Web site: http://www.ccsinfo.com

7. SIM340Z Modem User Manual
   Web site: http://www.roboeq.com/PDF/0502003.pdf

8. PCF8583 clock/calendar chip
   Web site: http://www.nxp.com

APPENDIX A

ASCII TABLE (7-BIT)

ASCII	Hex	Symbol
0	0	NUL
1	1	SOH
2	2	STX
3	3	ETX
4	4	EOT
5	5	ENQ
6	6	ACK
7	7	BEL
8	8	BS
9	9	TAB
10	A	LF
11	B	VT
12	C	FF
13	D	CR
14	E	SO
15	F	SI

ASCII	Hex	Symbol
16	10	DLE
17	11	DC1
18	12	DC2
19	13	DC3
20	14	DC4
21	15	NAK
22	16	SYN
23	17	ETB
24	18	CAN
25	19	EM
26	1A	SUB
27	1B	ESC
28	1C	FS
29	1D	GS
30	1E	RS
31	1F	US

ASCII	Hex	Symbol
32	20	(space)
33	21	!
34	22	"
35	23	#
36	24	$
37	25	%
38	26	&
39	27	'
40	28	(
41	29	)
42	2A	*
43	2B	+
44	2C	,
45	2D	-
46	2E	.
47	2F	/

ASCII	Hex	Symbol
48	30	0
49	31	1
50	32	2
51	33	3
52	34	4
53	35	5
54	36	6
55	37	7
56	38	8
57	39	9
58	3A	:
59	3B	;
60	3C	<
61	3D	=
62	3E	>
63	3F	?

ASCII	Hex	Symbol
64	40	@
65	41	A
66	42	B
67	43	C
68	44	D
69	45	E
70	46	F
71	47	G
72	48	H
73	49	I
74	4A	J
75	4B	K
76	4C	L
77	4D	M
78	4E	N
79	4F	O

ASCII	Hex	Symbol
80	50	P
81	51	Q
82	52	R
83	53	S
84	54	T
85	55	U
86	56	V
87	57	W
88	58	X
89	59	Y
90	5A	Z
91	5B	[
92	5C	\
93	5D	]
94	5E	^
95	5F	_

ASCII	Hex	Symbol
96	60	`
97	61	a
98	62	b
99	63	c
100	64	d
101	65	e
102	66	f
103	67	g
104	68	h
105	69	i
106	6A	j
107	6B	k
108	6C	l
109	6D	m
110	6E	n
111	6F	o

ASCII	Hex	Symbol
112	70	p
113	71	q
114	72	r
115	73	s
116	74	t
117	75	u
118	76	v
119	77	w
120	78	x
121	79	y
122	7A	z
123	7B	{
124	7C		
125	7D	}
126	7E	~
127	7F	.

APPENDIX B

SIM340 MODEM PIN ASSIGNMENTS AND PIN DEFINITIONS

Pin Number	Name	Description
1,2,3,4,5,6,7,8	VBAT	Supply voltage (3.4V – 4.5V)
9,10,11,12,13,14	GND	Supply ground
15	VRTC	I/O for RTC
16	SIM_PRESENCE	SIM card detection
17	VDD_EXT	2.93V supply output
18	DISP_DATA	Display interface
19	SIM_VDD	SIM card voltage
20	DISP_CLK	Display interface
21	SIM_DATA	SIM card data
22	DISP_CS	Display interface
23	SIM_CLK	SIM card clock
24	DISP_D/C	Display interface
25	SIM_RST	SIM card reset
26	DISP_RST	Display interface
27	KBC0	Keypad interface
28	DCD	Data carrier detect
29	KBC1	Keypad interface
30	NETLIGHT	Status indicator
31	KBC2	Keypad interface
32	GPIO0	I/O port
33	KBC3	Keypad interface
34	PWRKEY	Power on key
35	KBC4	Keypad interface
36	Buzzer	Buzzer interface
37	KBR0	Keypad interface
38	DTR	Data terminal ready
39	KBR1	Keypad interface
40	RXD	Receive data
41	KBR2	Keypad interface
42	TXD	Transmit data
43	KBR3	Keypad interface
44	RTS	Request to send
45	KBR4	Keypad interface
46	CTS	Clear to send
47	DBG_RXD	Debugging receive

48	RI	Ring indicator
49	DBG_TXD	Debugging transmit
50,51	AGND	Analog ground
52	ADC0	A/D converter input
53	MIC1P	Microphone 1 input (+)
54	SKP1P	Speaker 1 output (+)
55	MIC1N	Microphone 1 input (-)
56	SPKR1N	Speaker 1 output (-)
57	MIC2P	Microphone 2 input (+)
58	SPK2P	Speaker 2 output (+)
59	MIC2N	Microphone 2 input (-)
60	SPKR2N	Speaker 2 output (-)

APPENDIX C

SOME COMMONLY USED "AT" COMMANDS

Command   **Definition**

ATA   Answer a call
ATD   Call a number
ATE0   Disable echo
ATE1   Enable echo
ATH   Terminate a call
ATL   Set monitor speaker loudness
AT+CADC   Read A/D channel
AT+CALARM   Set alarm
AT+CHFA   Swap audio channels
AT+CMGD   Delete SMS message
AT+CMGF   Select SMS type
AT+CMGL   List SMS messages
AT+CMGR   Read SMS message
AT+CMGS   Send SMS message
AT+CMIC   Microphone gain level
AT+CPBR   Read phone book entries
AT+CPBW   Write phone book entry
AT+CPIN   Enter PIN
AT+CPWD   Change password
AT+CSCA   SMS service center no
AT+CSCS   Select character set
AT+CSMP   Set SMS text mode parameters
AT+CSQ   Request signal quality
AT+GMI   Request manufacturer ID
AT+GMM   Request model ID
AT+GMR   Request revision ID
AT+CMUT   Mute control
AT+COPS   Operator selection
AT+ECHO   Echo cancellation control
AT+SIDET   Side gain level

APPENDIX D

SMS DATA CODING SCHEME
(reprinted with the permission of ETSI, ref: 2010-001-07)

The TP-Data-Coding-Scheme field, defined in GSM 03.40, indicates the data coding scheme of the TP-UD field, and may indicate a message class. The octet is used according to a coding group which is indicated in bits 7..4. The octet is then coded as follows:

Coding Group Bits 7..4	Use of bits 3..0
00xx	General Data Coding indication Bits 5..0 indicate the following : Bit 5, if set to 0, indicates the text is uncompressed Bit 5, if set to 1, indicates the text is compressed using the GSM standard compression algorithm. (yet to be specified) Bit 4, if set to 0, indicates that bits 1 to 0 are reserved and have no message class meaning Bit 4, if set to 1, indicates that bits 1 to 0 have a message class meaning : Bit 1 Bit 0 Message Class 0 0 Class 0 0 1 Class 1 Default meaning: ME-specific. 1 0 Class 2 SIM specific message 1 1 Class 3 Default meaning: TE specific (see GSM TS 07.05) Bits 3 and 2 indicate the alphabet being used, as follows : Bit 3 Bit 2 Alphabet: 0 0 Default alphabet 0 1 8 bit 1 0 UCS2 (16bit) [10] 1 1 Reserved NOTE: The special case of bits 7..0 being 0000 0000 indicates the Default Alphabet as in Phase 2
0100..1011	Reserved coding groups
1100	Message Waiting Indication Group: Discard Message Bits 3..0 are coded exactly the same as Group 1101, however with bits 7..4 set to 1100 the mobile may discard the contents of the message, and only present the indication to the user.

1101	Message Waiting Indication Group: Store Message
	This Group allows an indication to be provided to the user about the status of types of message waiting on systems connected to the GSM PLMN. The mobile may present this indication as an icon on the screen, or other MMI indication. The mobile may take note of the Origination Address for messages in this group and group 1100. For each indication supported, the mobile may provide storage for the Origination Address which is to control the mobile indicator
	Text included in the user data is coded in the Default Alphabet.
	Where a message is received with bits 7..4 set to 1101, the mobile shall store the text of the SMS message in addition to setting the indication.
	    Bits 3 indicates Indication Sense:
	
	    Bit 3
	    0 Set Indication Inactive
	    1 Set Indication Active
	
	    Bit 2 is reserved, and set to 0
	
	    Bit 1 Bit 0 Indication Type:
	    0 0 Voicemail Message Waiting
	    0 1 Fax Message Waiting
	    1 0 Electronic Mail Message Waiting
	    1 1 Other Message Waiting*
	
	* Mobile manufacturers may implement the "Other Message Waiting" indication as an additional indication without specifying the meaning. The meaning of this indication is intended to be standardized in the future, so Operators should not make use of this indication until the standard for this indication is finalised.

1110	Message Waiting Indication Group. Store Message

The coding of bits 3..0 and functionality of this feature are the same as for the Message Waiting Indication Group above, (bits 7..4 set to 1101) with the exception that the text included in the user data is coded in the uncompressed UCS2 alphabet. |
| 1111 | Data coding/message class

Bit 3 is reserved, set to 0.

Bit 2   Message coding:
0   Default alphabet
1   8-bit data

Bit 1   Bit 0   Message Class:
0   0   Class 0
0   1   Class 1   default meaning: ME-specific.
1   0   Class 2   SIM-specific message
1   1   Class 3   default meaning: TE specific (see GSM TS 07.05) |

Default alphabet indicates that the TP-UD is coded from the 7-bit alphabet given in subclause 6.2.1. When this alphabet is used, the characters of the message are packed in octets as shown in subclause 6.1.2.1.1, and the message can consist of up to 160 characters. The default alphabet shall be supported by all MSs and SCs offering the service.

8-bit data indicates that the TP-UD has user-defined coding, and the message can consist of up to 140 octets.

UCS2 alphabet indicates that the TP-UD has a UCS2 [10] coded message, and the message can consist of up to 140 octets, i.e. up to 70 UCS2 characters.

When a message is compressed, the TP-UD consists of the default alphabet or UCS2 alphabet compressed message, and the compressed message itself can consist of up to 140 octets in total.

When a mobile terminated message is class 0 and the MS has the capability of displaying short messages, the MS shall display the message immediately and send an acknowledgement to the SC when the message has successfully reached the MS irrespective of whether there is memory available in the SIM or ME. The message shall not be automatically stored in the SIM or ME.

The ME may make provision through MMI for the user to selectively prevent the message from being displayed immediately.

If the ME is incapable of displaying short messages or if the immediate display of the message has been disabled through MMI then the ME shall treat the short message as though there was no message class, i.e. it will ignore bits 0 and 1 in the TP-DCS and normal rules for memory capacity exceeded shall apply.

When a mobile terminated message is Class 1, the MS shall send an acknowledgement to the SC when the message has successfully reached the MS and can be stored. The MS shall normally store the message in the ME by default, if that is possible, but otherwise the message may be stored elsewhere, e.g. in the SIM. The user may be able to override the default meaning and select their own routing.

When a mobile terminated message is Class 2 (SIM-specific), a phase 2 (or later) MS shall ensure that the message has been transferred to the SMS data field in the SIM before sending an acknowledgement to the SC. The MS shall return a "protocol error, unspecified" error message (see GSM TS 04.11) if the short message cannot be stored in the SIM and there is other short message storage available at the MS. If all the short message storage at the MS is already in use, the MS shall return "memory capacity exceeded".

APPENDIX E

SMS DEFAULT ALPHABET SET
(reprinted with the permission of ETSI, ref: 2010-001-07)

Bits per character: 7

SMS User Data Length meaning: Number of characters

CBS pad character: CR

Character table:

				b7	0	0	0	0	1	1	1	1
				b6	0	0	1	1	0	0	1	1
				b5	0	1	0	1	0	1	0	1
b4	b3	b2	b1		0	1	2	3	4	5	6	7
0	0	0	0	0	@	Δ	SP	0	¡	P	¿	p
0	0	0	1	1	£	_	!	1	A	Q	a	q
0	0	1	0	2	$	Φ	"	2	B	R	b	r
0	0	1	1	3	¥	Γ	#	3	C	S	c	s
0	1	0	0	4	è	Λ	¤	4	D	T	d	t
0	1	0	1	5	é	Ω	%	5	E	U	e	u
0	1	1	0	6	ù	Π	&	6	F	V	f	v
0	1	1	1	7	ì	Ψ	'	7	G	W	g	w
1	0	0	0	8	ò	Σ	(	8	H	X	h	x
1	0	0	1	9	Ç	Θ	)	9	I	Y	i	y
1	0	1	0	10	LF	Ξ	*	:	J	Z	j	z
1	0	1	1	11	Ø	1)	+	;	K	Ä	k	ä
1	1	0	0	12	ø	Æ	,	<	L	Ö	l	ö
1	1	0	1	13	CR	æ	-	=	M	Ñ	m	ñ
1	1	1	0	14	Å	ß	.	>	N	Ü	n	ü
1	1	1	1	15	å	É	/	?	O	§	o	à

INDEX

#

#define, **56**
#elif, **56**
#endif, **56**
#error, **56**
#if, **56**
#ifdef, **56**
#ifndef, **56**
#include, **57**
#undef, **56**

A

addition, **53**
analogue comparator, **22**
Array pointers, **48**
Arrays, **46**
arithmetic operators, **53**
ASCII table, **186**
AT command, **124**
ATE0, **126**
ATE1, **126**
ATH, **141**
ATS3, **125**
auto decrement, **53**
auto increment, **53**

B

BCD, **177**
brown out detector, **20**

C

CAN, 21,**23**
case sensitivity, **38**
CBAND, **145**
CCVM, **145**
CHFA, **140**
CIMI, **146**
CISC, **25**
Clock, **19**
Clock configuration, **29**
CLTS, **145**
cellular network, **113**

CMGDA, **128**
CMGF, **128**
CMGS, **128**
CMGW, **128**
CMSS, **128**
CMUT, **147**
CNUM, **146**
comments, **37**
constants, **42**
CPIN, **128**
CRSL, **146**
Crystal, **30,31**
CSCS, **129**
CSMP, **129**
CTS, **121**
current sink, **27**
current source, **23**

D

DA, **135**
DCD, **121**
Delay_Ms, **80**
division, **53**
do statement, **65**
double, **42**
DTR, **121**

E

EasyPIC 5, **149**
EEPROM, *18*
else, **58**
embedded controller, **15**
enddo statement, **76**
Enumarated constants, **44**
enumarated vaiable, **45**
EPROM, **17**
Escape sequences, **44**
Ethernet, **24**
External reset, **34**

F

float, **40**
for statement, **61**

G

Goto, **66**
GSM operator, **113**
GSM system, **113**

H

Harvard, **25**
HD44780, **95**
HS_OSC, **81**
HyperTerminal, **110**

I

ICSP, **24**
If, **58**
INTCON, **154**
Interrupt, **20**

L

LCD, **97**
Lcd_Chr, **97**
Lcd_Chr_Cp, **97**
Lcd_Cmd, **97**
Lcd_Config, **97**
Lcd_Init, **97**
Lcd_Out, **97**
Lcd_Out_Cp, **97**
LED flashing, **36**
LM35DZ, **157**
logical operator, **55**
long double, **42**
longer delay, **85**
low power operation, **23,118**
LVP_OFF, **81**

M

MAX232, **104**
message encoding, **134**
modem card, **120**
MR, **135**
multiplication, **53**

O

Operators, **53**

P

PCF8583, **176,177**

PDL, **74**
PDU, **134**
PIC16F887, **26**
PIC18F, *19*
PID, **129**
Pointers, 47
pre-processor, **53,55**
PROM, **17**
push button, **90**
PWM, **24**

R

RAM, **17**
Real time, **22**
relational operator, **55**
Repeat statement, **75**
Reserved names, *39*
Reset, **32**
RI, **121**
RISC, *25*
ROM, **17**
RS232, *102*
RS232 connector, **104**
RTC chip, **176**
RTS, **121**
RXD, **121**

S

SCA, **135**
selection, **57**
sending SMS, **128,134**
serial communication, 102
signed char, **40**
signed int, **40**
signed long, **40**
SIM340Z function, **119**
SIM340Z modem, **118**
SIM card, **117**
SIM card holder, **117**
sizeof, **51**
Sleep mode, **22**
Smart GSM/GPRS board, **120**
SMS, **120,124**
STATUS register, **27,177**
String constants, **44**
Structures, **52**
subtraction, **53**
supply voltage, **30**
switch statement, **59**

T

temperature sensor, **57**
text mode, **128**
Timer, **19,178**
TNO, **135**
TXD, **148**
typedef, **52**

U

UD, **135**
UDL, **135**
unsigned char, **40**
unsigned int, **40**
unsigned long, **40**
USART, **104**
USB, **23**

V

Variable names, **39**
Variable Types, **40**
VP, **129**
VPP, **116**

W

Watchdog, **20**
WDT_OFF, **81**
while statement, **63**
white space, **138**

Z

ZigBee, **24**

VDM publishing house ltd.

Scientific Publishing House
offers
free of charge publication

of current academic research papers, Bachelor´s Theses, Master's Theses, Dissertations or Scientific Monographs

If you have written a thesis which satisfies high content as well as formal demands, and you are interested in a remunerated publication of your work, please send an e-mail with some initial information about yourself and your work to *info@vdm-publishing-house.com*.

Our editorial office will get in touch with you shortly.

VDM Publishing House Ltd.
Meldrum Court 17.
Beau Bassin
Mauritius
www.vdm-publishing-house.com

Made in the USA
Lexington, KY
26 February 2011